America's Electric Nazca

*

MEGALITHIC MARFA, TEXAS

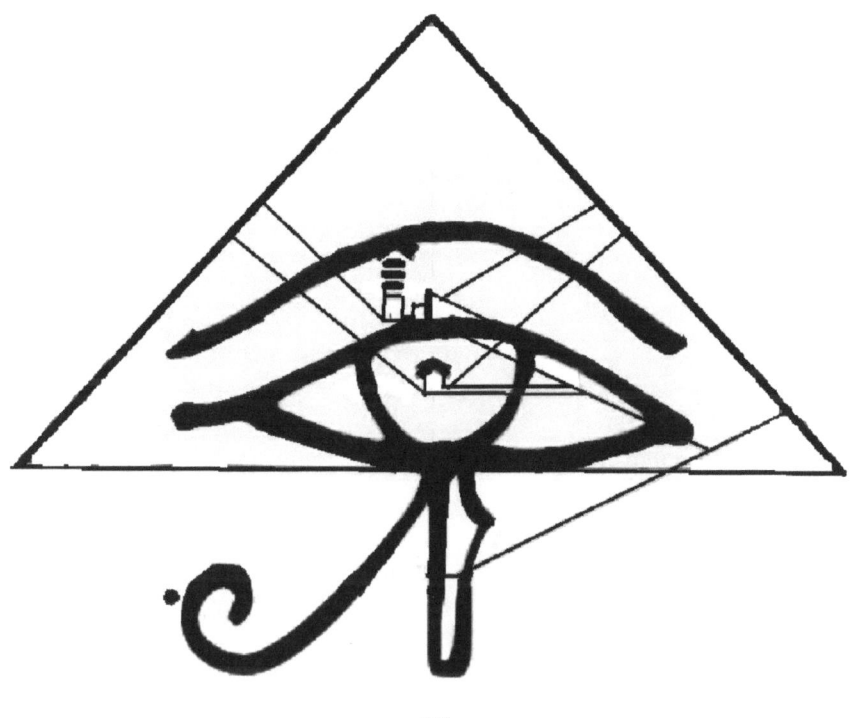

BY

Dan Hoquist

New Rosetta Discoveries Track Ancients, Their Paths,

Their Treasures, And Connect The Dots

ACKNOWLEDGEMENTS

Gerald Clark, MSEE PSI http://www.geraldclark77.com Author: The Anunnnaki of Nibiru- Mankind's Forgotten

Creators, Enslavers, Destroyers, Saviors, and Hidden Architects of the...New World Order

and:

The 7th Planet Mercury Rising

**

Edson C. Hendricks

San Diego, CA 92109-2357, USA

marfa@edh.net

**

Andis Kaulins

http://andiskaulins.com

B.A. University of Nebraska

J.D. Stanford University Law School

Former Lecturer in Anglo-American Law, FFA, Trier Law School

Alumnus Associate of Paul, Weiss, Rifkind, Wharton & Garrison, NY

Author (most recent) - Season Of Birth, Marriage And Profession

THANKS TO:

Lewis Travland (Pilot)

Travland Helicopter Intnl. Inc.

PO Box 2019 Alpine, Tx 79830

Ph # 432-837-1848

Additional Aviation Resources

 Near Marfa, Tx:

Cade Woodward (Pilot)

Woodward Aviation

Ph# 432-294-2629

Marfa Gliders - Restrictions Apply

www.FLYGLIDERS.com

**

SPECIAL THANKS TO:

Google Earth, Google Inc.

For Consistantly Providing High Definition

Satellite Imagery For Public Education

All Guidelines and Terms of Use have been Followed

TABLE OF CONTENTS

Preface

Disclosure. Anybody who is interested in alternative explanations to 'Conventional', e called, Scientific Theories of History, will eventually come-up against the wall, of Denial. Even armed with the other, 'D', word; Discovery, and all the Tools of Research and Artifact, Academia, combined with the weight of their Credentials, will simply refuse to include even discuss, the facts presented as Discovery. As those Credentialed Wizards of Oz, have been known to even change Recorded History, to satisfy Political, Religious and the Po Elite. Even to the point of Malicious destruction of whole Empires, which are now nothing more than Myth, in some lost Textbook. I would direct People to research the Lost King of Africa, whose true History includes Megaliths, the Builders and Structures now stand ignored! Or to the whole Empire of Tartaria, Russia, in where the Romanoffs were assisted to destroy Tartaria, the Coat of Arms, of which, depicted an OWL. After the destruction, this same Owl, starts showing-up on other so-called Noble's, Coats of Arms, indicating the assistance in this Murder-most-foul! Accordingly, the Romanoffs RE-NAMED every Hill, Mountain, River and Valley! Then, the Romanoffs, FELL to the same Treachery, by the forces that made the Plans, erasing all Evidence (so they thought). But the Tartars Coat of Arm fortunately remained as a reminder that the Truth will, eventually, be found. My personal favorite European Researcher is; the Newearth Youtube Channel, and Megaliths.org .

More to the Point, I have found out, that you can make the most incredible Discovery be actively ignored. For instance: I have now contacted all Newspapers, Radio, in the A Marfa, Texas. This happens to be the single Largest Archaeological Discovery in North America. And that, is not an exaggeration, as the area, in question, measures a Land Mass of a Hundred Square Miles, or more. Just to the Date of this writing, I have found more than 50 (fifty), Geoglyphs. Any ONE of which, if found on Mars, would be the only Story Media News, and would already have a count of several Million hits, on Youtube!

Google earth remains the only reliable source of Satellite imagery, but we get better Pictures from Mars, than we do with a hundred times more Satellites in Orbit around Earth!

Which is Why, I feel compelled to write this Book...because I now have Credentials of own; I am now an Author. Albeit, I've always considered myself a great Reader, this is a new experience I've been pressured into. The Journey of Discovery, has been almost a year, in the making, and the evidence, is the Path I take.

Which brings me to Marfa, Texas. I never thought I would be going back to Texas. I lived in Dallas, for almost 4 years, and even as a young Man, I couldn't take the Texas Winters. So, I left my Brother there, and Hitched my way back to Southern California. Yes, I've got Ex'es in Texas. It's just too bad they are not the people in charge; I have contacted, to date, Texas State Univ., Archaeology and Anthropology Depts, Texas Historical Society, 2 different archaeologists in Presidio County (where Marfa is), even San Antonio MUFON (and many more), to deafening Silence!

Maybe Marfa, TX is just not Sexy? BUT, it may explain...everything! And why not? Texas is a BIG State!

Until the Public is informed, until this Discovery is actively being Researched, I remain: The Man Who Ran Around Texas, Until They Called it: Troy!

1. EUREKA

Skeptical Attitude, or Temper, Doubt

Doubt or Unbelief in regards to Religion, especially Christianity

The Doctrines or Beliefs of Philosophical Skeptics; Universal Doubt

I would add that, Skeptics must be able to Test what they Doubt, as Good Scientists will run Experiments that don't agree with their Theories. Denial is NOT skepticism.

I will go even farther; a 'True' Skeptic, is able to Set-Aside their Doubt, and ACCEP Evidentiary Results, NO MATTER where the Evidence Leads...

Like the Fictional Character; Normal Norman, I have no Degrees, past Exploits of Greatness, Banners or Trophies. But I AM a Journeyman Sheet metalworker, or was, 22yrs in Construction, work dried-up in 1992, and I was only able to Log, One Month, in Trade.

So, I considered both, going to School for Nursing, and taking the Postal Exam. I tested in the high 90%, at both opportunities, but the US Postal Service called first, and they had a Pension. I went from $25/hr, to $6/hr, with only a 3 Month guarantee, worked my Butt-off, and was picked-up at a 1 1/2yr "Transitional Carrier", and much better pay, still no Benefits.

I became a Rural Mail Carrier, the difference being; you worked more as a 'Sub- Contractor' for US Mail Delivery, and they didn't give you Uniforms...OH, and you Delivered the Mail from the Passenger Seat of your PERSONAL vehicle! Couple Sets of Brakes and Tires per Year! Of course the USPS would compensate us for Mileage, but most Americans are unaware of this kind of Postal Delivery.

I was Injured, on-the-Job, in 2005, but the USPS has 'Light-Duty', with which they try to get you SO Bored, or subtly Harrassed, that you will eventually go back to your Work, Hurt or NOT, or you can always Quit. I'm just as sure they don't mind EITHER choice, they just want you if you're REALLY Injured!

I got worse, and Medically Retired in 2008.
If you need some Credentials, or a PHD, to accept Information or New Discoveries I suggest you look into a Garage Electrician, whose Name is John Hutchison. I've worked several PHD's in my 60yrs, the LAST one Delivered Mail, and NOT that well. You would be well-reminded that the Dead Sea Scrolls and Gobekli-Tepi, were found by Goat-Herders, and the story of Troy's Discovery, made by a Week-end Amatuer Archaeologist; who Ran around a Hill.

I like to believe I have a very 'Mechanical' Mind, and Worked in Supervisoral Roles, Construction and Postal, at times was Lead-Man, over 25, or more Workers. I seem to have good Linear sense; I can look at a Place, and judge Distance very well and had a knack making Material lists. This Book is NOT about me, though, but about what I found, and their Implications. I was Widowed in December, 2008, when my Wife of 30yrs Passed. As a part of Grief relief, I wanted to make some Changes to the House, that more reflected MY character designed some Landscape; a 'Zen' Garden (dry, because of Calif Drought), a 3-Arch 'Shinto' Patio Cover (Health, Fortune, Love), and a Yin-Yang symbol done as a Concrete Pad.

But the Centerpiece of my Zen Garden, is going to be a Scale Model Stonehenge Replica done in Crystal, & colored Glass, on a Pexiglass Base, so I can LIGHT the WHOLE thing with Multi-colored LEDs! This will be on the Top of a Spanish-

Moss Covered 'Island' with assorted 'Islands' throughout the Garden.

Next to the Garden will be a short Deck area, with a Fire-pit, to allow contemplation of a Mt Fuji Mural, and the Zen Garden.
Because I am disabled, I just do the Design, and Hire Friends-with-Talent. My Muralpartially done, at this point, and I might attempt to complete the Scene, my Artist is Tempermental. But, I think everything will be complete sometime in 2017.

2016 looks to be a busy Year, because of a Discovery I made, while doing Research on Stonehenge.

Of Course, any Research starts with a Blueprint of Stonehenge, and I found one Online that showed ALL the Stones, as if the Structure were new. But any Research would have to include Satellite Images, of the real thing, so, Google Earth was a MUST. (Note: although I don't get paid for Advertising; only Google Earth provides consistent HD Satellite Imagery).

My Interests include the Ancient 'Alien' Theories of Erik Von Danniken, and Zechar Sitchen, and I just happened to be watching an old Lecture done by Mr. Sitchen, when he called the Anunnaki, "The Harvesters Of the Sun".

As a Side-Note: I have a tendency to have the Television playing some Movie, while Listening to either Radio, or some Youtube Lecture, so that I can work on something else.

And I was working on my Scale Stonehenge, when I heard Mr. Sitchen utter that Phrase, I saw that Stonehenge was a TOOL, they USED FOR HARVESTING THE SUN!

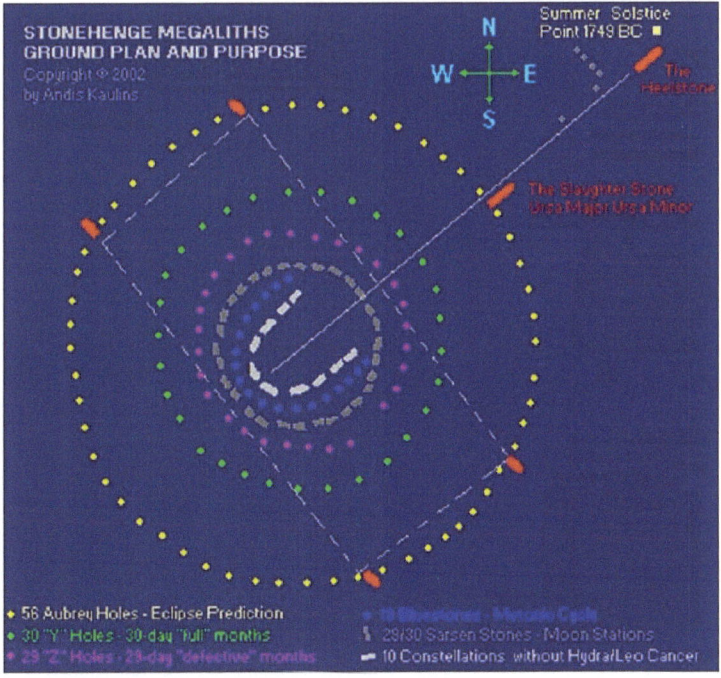

STONEHENGE MEGALITHS
GROUND PLAN AND PURPOSE
Copyright © 2002
by Andis Kaulins

N
W — E
S

Summer Solstice
Point 1749 BC ■

The
Heelstone

The Slaughter Stone
area Major Ursa Minor

◆ 56 Aubrey Holes - Eclipse Prediction
◆ 30 "Y" Holes - 30-day "full" months
◆ 29 "Z" Holes - 29-day "defective" months

◆ 19 Bluestones - Metonic Cycle
↕ 29/30 Sarsen Stones - Moon Stations
� 10 Constellations without Hydra/Leo Cancer

If you will notice, the Center, 'U' Section, seems to cup the Summer Solstice Sun, when the Sun is 'Ripe', or when the Sun gives the Earth it's most Daylight, it does not allow the light to escape, but ENDS at the Altar.

My Mechanical Mind just can't seem to stop asking 'WHY?', so, the connection I made was ,' how would I make a Tool to Harvest the Sun?', and, 'when would I harvest, to get it at it's Peak?'

The answers seemed obvious, with the amazing Math involved with Stonehenge, you wouldn't need to know the Science, you could teach Monkeys to value the Solstice, and to place any 'Pieces of the Sun'. Which is, the name that many Tribal, and Ancient People called GOLD...on the Altar. That's when I found Google Earth had a 'Layer' function, in their Toolbox.

2. DRAWING LINES

Google Earth is a MARVELOUS Tool.

Google Earth HAS WONDERFUL Tools to work with!

Fig 2

BLUEPRINT FOR STONEHENGE OVERLAYED AT STONEHENGE, UNITED KINGDOM

I first used Google's Layer Tool on the real Satellite Image of Stonehenge, UK. It seemed only natural to follow the Summer Solstice Line, given in the Blueprint. Then I used the Line Tool, and found how precise Go Earth IS; when you STOP and Save a piece of Line, your attempt to Connect your piece MUST be e or it will look 'off'.

It also told me a lot about the People who built these Shrines. They were exacting, these Solstice Lines were precise, by inches!

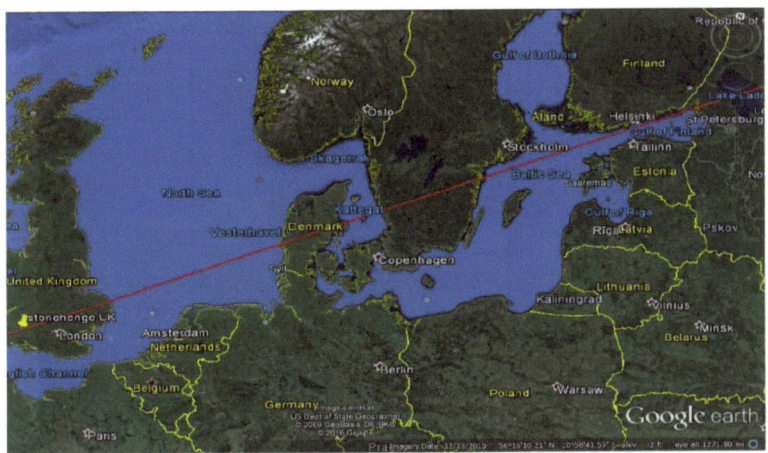

Fig 3

SOLSITCE LINE FROM STONEHENGE PASSES THROUGH SAINT
PETERSBURG, RUSSIA

Stonehenge Solstice Line is aimed straight for St Petersburg, Russia,
then through the Gobi Desert, China, and Taiwan.

I would suggest that any, who think this is not significant, go Online
and view the Saint Petersburg videos, part of a great Youtube
Educational Series called; 'The Survivors', developed by a European
Researcher, under the Newearth Youtube Channel.

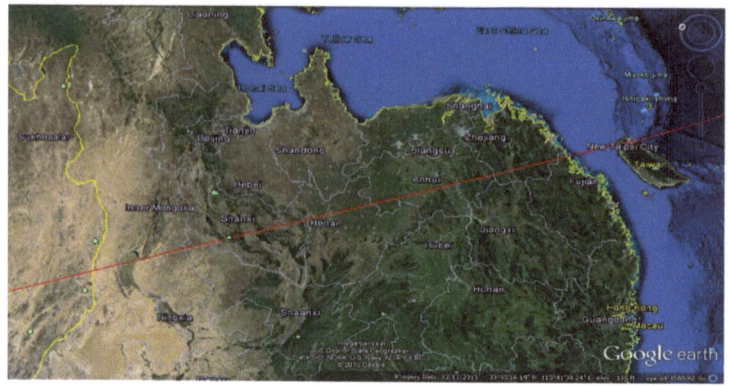

Fig 4

LINE CONTINUES THROUGH MONGOLIA, CHINA & TAIWAN

From Taiwan, China, this Solstice Line cuts through the heart of New Guinea, New Zealand, where there is apparent knowledge of this line, as it goes through the CENTER of Christchurch, and through a 'too round' are (like an Artificial Land mass designed to be another 'Stonehenge-like' Shrine,

It seems that the people using these Lines, used Land Masses, as well as Shrines, for some kind of Navigational Marker, all at a Time in History, Archaeologists and Anthropologists assure us, Humans did not fly!

Christchurch to Santiago, Chile, this line exists South America through Brazil, at the Baia De Marajo, near the Mouth of the Amazon River.

The Mouth of the Amazon River has been the site of a fairly new find that shows a population of several Million, living, in what is now, dense jungle.

Traces of massive tracts of Land that were being used for Agriculture have been found after recent De-forestation of the Amazon Basin. Apparently, we are 'saving the Rainforest', at the cost of 'losing' the Archaeological Sites, under the existing growth.

This also gives us a Timeframe, which can be found in some 'Tree-rings', yet to be counted, but from a layman's point of view, it's got to be Thousands of years!

Fig 6

STONEHENGE LINE PASSING THROUGH SOUTH AMERICA @ SANTIAGO, CHILE EXITING THE MOUTH OF THE AMAZON RIVER

And, back again, to the Stonehenge Circle, itself. That this Solstice Line connected with ANYTHING was impressive. That this Line connected to SO MANY places of 'import', went beyond any Statistical Math I knew of. If you were to think of it in Lottery terms: you would WIN, literally, EVERY time you PLAYED!

On a 'Lark', I decided to try Google's Layer function on the area I live in; the San Gorgino Pass, Southern California, because I had thought that Mount San Jacinto, could easily be a 'marker' for the Winter Equinox. Indians I know had told me, since childhood, that San Jacinto, "didn't belong, where it stands", and that, "the Mountain is Hollow"!

Indeed, when looked-at, from a good Topographical Map, the Mountain seems out-of-place from the Pacific Crest, and could easily be mistaken for a 'blasted' Pyramid, albeit...almost 12,000ft, at the Peak!

Fig 7

MOUNT SAN JACINTO, SOUTHERN CALIFORNIA

So I re-sized the Stonehenge Blueprint and noted Local landmarks that co-incided, but not quite. After a couple of Layers, I noticed that I was coming closer to an Apex, or meeting point...at Santa Catalina Island!

3. GIANTS

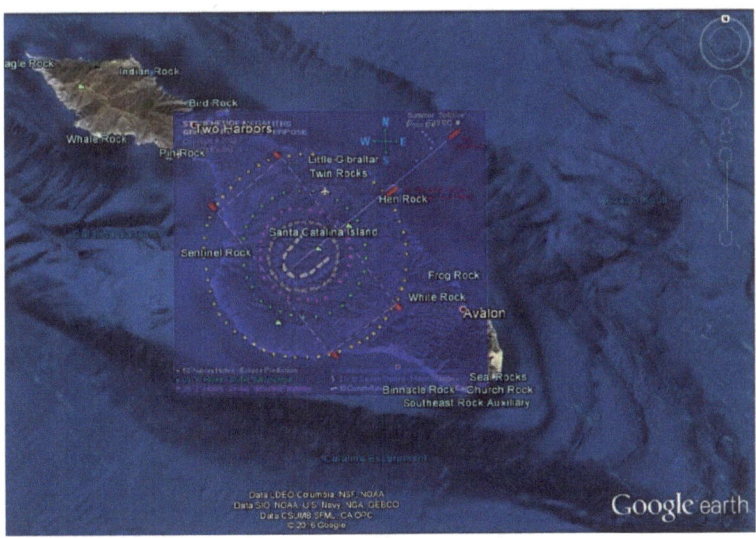

FIG 8

SANTA CATALINA ISLAND OVERLAYED WITH SONEHENGE

Now you might see a Pattern emerging, THIS is how my Brain works; I follow the evidence, I do NOT make the Evidence.

I do not 'dismiss' anything, without the willingness to try the 'thing', firstly, neither will I accept blindly, without the necessary Study, to Verify.

When I started to study Santa Catalina Island, I found that it was known as; Stonehenge LONG BEFORE any Archaeologic Study. Which was eventually done in 1920, when they found GIANTS!

Ralph Glidden was digging on Catalina Island in the Gulf of California between 1919 and 1928. He found, according to newspaper articles, numerous skeletons 7 to 10 foot tall. But where are those skeletons today? Was it just a publicity stunt - or did he find the remains of a lost race of human giants?

Santa Catalina Island, most often just called Catalina Island, is one of the Channels Islands off the coast of California in USA, just an hour by ferry from Long Beach in Los Angeles city. It is 22 miles (35 kilometers) miles long and 8 miles (13 km) across at it widest.

The island is quite rocky and was previously called; Stonehenge Island, by the Locals, the highest point is 2,097 feet (639 m). The Channel Islands has the earliest evidence for seafaring in the Americas.

Ralph Glidden was an amateur archaeologist who uncovered ancient burial sites on Catalina Island from 1919 to 1928: it is said that he excavated more than 800 grave-sites from about 100 individual locations around the island. In addition to finding thousands of artifacts he also dug up almost 4000 human skeletons.

He claimed that there once had lived an ancient race of tall and fair haired Indians on Catalina Island and the adjacent islands. Most of all the male adults were around 7 feet (2.20 m) in height, and largest skeleton he found was 9 feet 2 inches (2.80 m) tall.

Glidden lost his sponsor after digging for almost 10 years and the general opinion is that he then was bluffing about finding giant skeletons to create interest and make money.

Well, might be, but I'm not so sure. Ralph Glidden was not the first to find a giant skeleton on Catalina Island. According to the Pittsburgh Press July 20, 1913 and also Daily Telegraph on July 26; a German naturalist Dr. A.W Furstenan unearthed an 8 foot skeleton on the island. The skeleton was found artefacts such as mortars, pestles and arrowheads; all different from ordinary Indian relics. Plus a strange, flat stone, bearing unknown symbols.

Furstenan had while in Mexico heard a legend of a noble race of giants that had lived on Catalina Island long before the white man had arrived - and travelled to the island to investigate.

He found the skeleton along Avalon Bay, in black, hard sand showing signs of burning. Most of the bones crumbled to dust when they came up in the air; only the skull, jawbone and bones of one foot remained.

Also the other islands.

It is not only on Catalina Island it seems to have been found giants skeletons, the reports of findings also on other of the Channel Islands. According to newspaper artic les, Santa Rosa Island was the site of a dig in 1959, and they discovered several skeletons more than 7 feet tall. The tops of the skulls were painted red. The skulls were said to be of "primitive", "otamid" type: and also most often described as having sloped forehead, pronounced brow, robust bones, powerful jaw, and a so called "inca bone" at the back of the skull. were also said to have double rows of teeth.

The skeletons were found at an Indian cemetery that contained abalone shells they were dated to be more than 7.000 years old. According to the book "The Native Races o Pacific States of North America" (a five-volume description of indigenous ethnic group Hubert Howe Bancroft a Mr. Taylor from San Buenaventura stayed at Santa Rosa Island 1861 and he often came across skeletons of Indians in caves with double rows of teeth Santa Rosa Island is the second largest of the Channel Islands of California, about 150 northwest of Catalina Island.

Also on San Nicolas Island west of Catalina there shall have been found large skeleton In 1897 a party of relic hunters stayed three weeks on the barren island and Newark D Advocate tells about "Bones of a Giant Race on San Nicolas Island". The party found 87 buried in the sand but only three were secured entirely.

Trace of Giants Found in Desert

LOS ANGELES, Aug. 4 (AP)—A retired Ohio doctor has discovered relics of an ancient civilization, whose men were 8 or 9 feet tall, in the Colorado desert near the Arizona-Nevada-California line, an associate said today.

Howard E. Hill, of Los Angeles, speaking before the Transportation Club, disclosed that several well-preserved mummies were taken yesterday from caverns in an area roughly 180 miles square, extending through much of southern Nevada from Death Valley, Calif., across the Colorado River into Arizona.

Hill said the discoverer is Dr. F. Bruce Russell, retired Cincinnati physician, who stumbled on the first of several tunnels in 1931, soon after coming West and deciding to try mining for his health.

MUMMIES FOUND

Not until this year, however, did Dr. Russell go into the situation thoroughly, Hill told the luncheon. With Dr. Daniel S. Bovee, of Los Angeles—who with his father helped open up New Mexico's cliff dwellings—Dr. Russell has found mummified remains, together with

SO, now I had Anunnaki, Stonehenge, Solstice Lines, Gold and Giants, I did the natural thing; I made my first Bad Video, 'Giants, How to Locate Them, AND their Gold' (https://youtu.be/H-bMHhuVqEY) . I believe that the Solstice Lines were of UTMOST importance, to these People, that the Anunnaki USED the Lines, to somehow Collect the 'Pieces of the Sun', offerings, FROM them!

Less than Two Weeks after i posted that Video, the News reported that someone had found a $5,000,000 dollar Mine, where, was not noted. I also found out that the OTHER Mountain area, on the North side of the 'Pass', are the San Bernardino Mountains, which, in the 1930s had been found to be 'hollow', and a couple of Men walked-out with about 200lbs of Gold.

Fig 11

CAVERNS INSIDE SAN GORGONIO MOUTAINS

I would suggest, to anyone looking for gold, they can re-purpose this information to better assist in discovery. But, to me, I believe that Gold is Addictive AND Poisonous, and any 'real' value it has, is simply for Adornment, ME? I think I might be Allergic to it, as I get Bone Spurs, and sores whenever it touches my skin, if don't too long. I don't "Accessorize"!

4. FOLLOWING LINES

I was not ready to start 'Zipping' around the World, yet. I
drew a line from Santa Island and found that it passed
directly over the Grand Canyon. Feeling inspired, I thought,
'if Southern California has a Stonehenge, then why not the
United States?', or more accurately, North America?

I resized the Stonehenge Blueprint, and used the Grand
Canyon as an 'Altar'. The Summer Solstice Line taken from
Santa Catalina EXACTLY matched the Line of Solstice,
the Grand Canyon! Like a Kitten to a Ball of Yarn, I was
HOOKED. The Solstice Line from Catalina HAD to end,
somewhere.

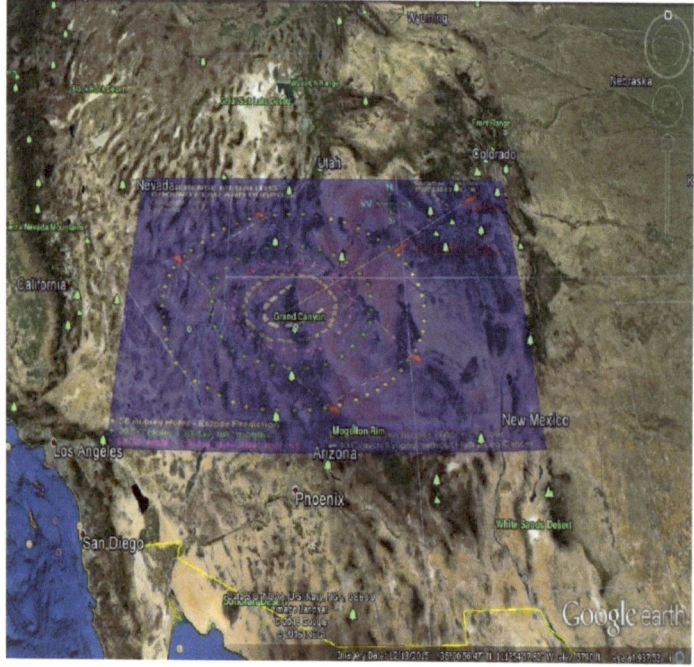

Fig 12

THE GRAND CANYON AS ALTAR OF STONEHENGE

It does, at the Physical Furthest Place Where Land Mass Meets The Atlantic Ocean, Nova Scotia, where an Island, stands, just off of the Coast, that just Happens To Look Like A Giant Footprint! My Michael Tellinger moment.

Fig 13

SOLSTICE LINE FROM SANTA CATALINA TO NOVA SCOTIA USING GRAND CANYON AS STONEHENGE ALTAR

Fig 14

GIANT FOOTPRINT ISLAND OFF NOVA SCOTIA SOLSTICE LINE FROM CATALINA

Mr. Tellinger tells of finding a Giant Footprint, that began his Out Of Place Archaeology, and chase of the Race of People, seemingly OLD, before Modern man stopped dragging his Knuckles.

Those who know the area, will be more aware of a Famous UFO incident, that occurred in a place called; Shag Harbor, and you will find that this Line is very close to that area.

I had also noted, the Solstice Line also Matched a Map of Underground Tunnels, and Military Bases, that an Engineer, Named, Phil Schnieder, used to show, when he Lectured.

Fig x

Map Of Tunnels Under The United States before its new use of material does NOT imply Endorsement

For those who don't follow Paranormal News; there is purportedly, an Entrance to a Huge Tunnel area, near Malibu Beach, that stretches across the North American Continent. I believe this Solstice Line is another Validation of the Late Mr Schnieder's work, as his work, in turn, validates the importance of the Solstice Connections.

I thought this Solstice line was aimed at the UK Stonehenge but it aimed at a, seemingly, MUCH larger Footprint...the Island Continent of Madagascar itself!

I understand any Scepticism, contemplating Construction done on this kind of Scale, borders on impossibility!

But the Sumerian History actually details this kind of Terraforming, and Hindu Vedas are incredibly descriptive of Machinery, even Robots. I have found evidence that this kind of Construction HAS BEEN DONE!

Fig 15

MADAGASCAR JUST A GIANT STEP FROM AFRICA

This Line continued through Australia, passing directly through the Fiji Islands, where the Lost continent of MU was said to lie.

Indeed, looking at this Satellite image, you can see signs that the Ocean Seas were at lower levels, as there are traces of Coastlines, even Harbor, and a River! Although the age must stretch far into the Past, Myth still provides more information than modern Human History reveals.

Again, I feel I must mention: Denial is NOT Skepticism, and refusal of Artifactual Evidence is NOT Science!

From the South Pacific Ocean, here this Line connects, again, to Santa Catalina. My next step, was to follow the Footsteps of the Giants, but WHERE to start? Notifying Mr Tellinger, seemed like a good idea, so I made another bad video, and entitled it, "For Michael Tellinger, South African Giants"; (https://youtu.be/y9-Y_clru-8) In this video, I did Stonehenge Layer, direcly OVER the 'Adam's Calender' area, that Mr Tellinger has famously researched.

Fig 16

LINE PASSING THROUGH THE HEEL OF MADAGASCAR

I naturally thought Mr Tellinger might be interested in a slightly larger Giant footprint, one that connected DIRECTLY to the Adam's Calendar.

It was just around this time I was searching Youtube for anything related to Giants, and Ancient, hidden History, when I came across the Newearth Channel, and their 'Survivor' series. Not only, is this incredibly well done, it gave me a European look at THEIR hidden Histories. I would suggest viewing the whole series, the Russian Woman, who Narrates, has a wonderful voice, and adds an Artistic flavor...WELL DONE!

Fig 17

SOLSTICE LINE USING ADAMS CALENDER, AFRICA AS STONEHENGE

According to the Sumerian Chronicles, and Modern 'History', Mankind is said to have come from Africa, to then encompass the World. But, if Cuneiform stands as the World's oldest writing, then, arguably, the Vedas, are Second, ONLY to this, and the Solstice Line coming FROM Adam's Calendar pointed DIRECTLY at India!

NOTE: This Solstice Line crosses the Giant 'Foot' of Madagascar. It continues from India, crosses into China, at Myanmar, then crosses DIRECTLY over Honshu, Japan.

Again, those who have followed the History of Unidentified Flying Objects might be aware of the Ancient Tale of a Downed Flying Craft, recovered by the People of Honshu, and the Goddess who gave them Culture and Technology as "thanks".

Fig 18

ADAM'S CLANDER SOLSTICE CUTS ACROSS MADAGASCAR

From Japan, the Solstice Line from Adam's Calendar travels to Chile, again, near to Santiago, across South America, it bisects Buenos Aries, and back to Africa, directly through the Gold Mines of South Africa, and back to the Calendar.

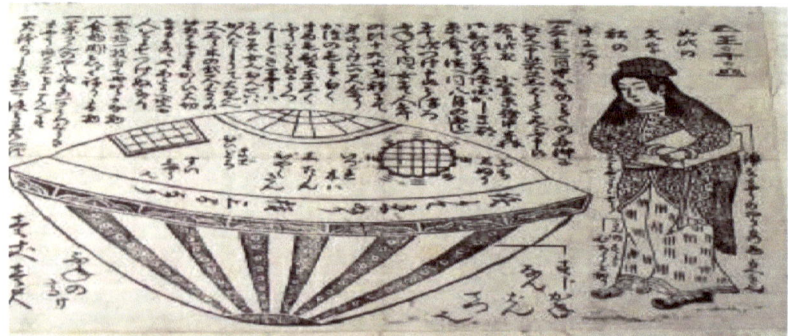

Fig 19

HISTORICAL SCROLL OF ANCIENT JAPANEESE INCIDENT WITH FLYING CRAFT

Santiago, Chile seems more and more, a Vertex point, and the opposite side exits the Continent at another River Mouth and Harbor. It's doubtful if those Residents of Buenos Aries even realize their Ancient History.

Fig 20

SOLSTICE LINE THROUGH CHILE, EXITS AT BUENAS ARIES

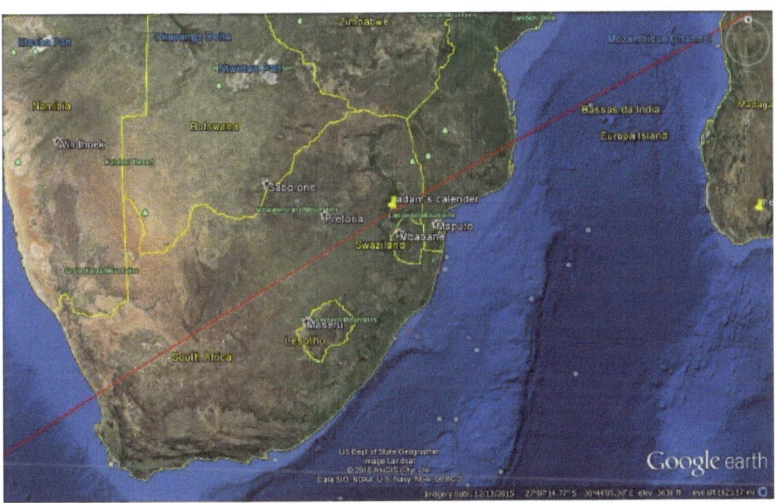

Fig 21

FROM SOUTH AMERICA TO ADAMS CALANDER THROUGH SOUTH AFRICA

With 'Coincidence' no longer an Evidentiary possibility, I was forced with the Conclusion, that if our History COMES from Africa, then the Giants walked a Solstice 'Path'.

5. WALKING THE SOLSTICE PATH

My next Layering, I had decided to try Machu Picchu, not because it's round or because of it's obvious 'Fame'. Little did I know, but I was going to, what can only be; the PRIMAL, or FIRST, Solstice Path. Yes, the term, 'Path' had replaced 'Line', as I was now sure that the People who had passed, through the Land, did so, once they had established a 'Shrine', using the amazing and 'weird' math, inherent in Stonehenge-like construction, guidance or Navigation, around the World.

I was totally unprepared for the Importance of the Path, and it's Historic value.

Fig 22

MACHU PICCHU WITH SONEHENGE OVERLAY

The Solstice Path from Macchu Picchu travels directly through the Mouth of the Amazon River. This marks another Vertex from Santiago, Chile, obviously not coincidental!

Across the Atlantic Ocean, the Path crosses through Cape Verde, but as it crosses Africa, through the Sahara Desert, it passes directly through the 'pupil' of the eye of Horus! At least, that's the only Name I can come-up with for a CONSTRUCTED area, INTENDED TO LOOK LIKE AN EYE! It measures approx 170mi long, by 50miles wide, at the 'pupil'. Archaeologists have only recently confirmed it's Artificial Nature: each concentric 'ring' of the Eye, is a different kind of Soil!

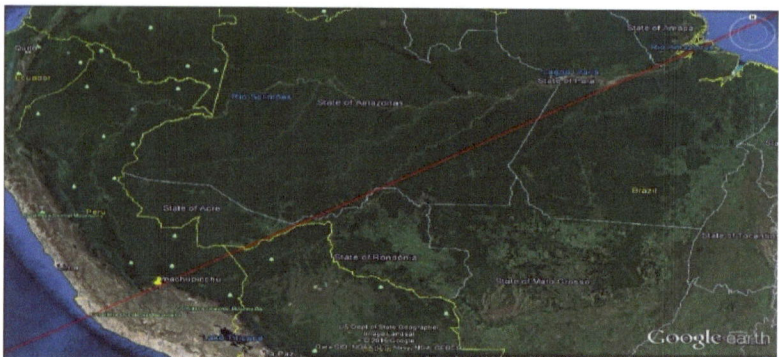

Fig 23 MACHU PICCHU SOLSTICE LINE EXISTS SOUTH AMERICA AT MOUTH OF AMAZON RIVER

Apparently the Soil was brought in purposely, to get the Effect of Coloring desired. OH, and it's OBVIOUSLY intended to be seen FROM ALTITUDE!

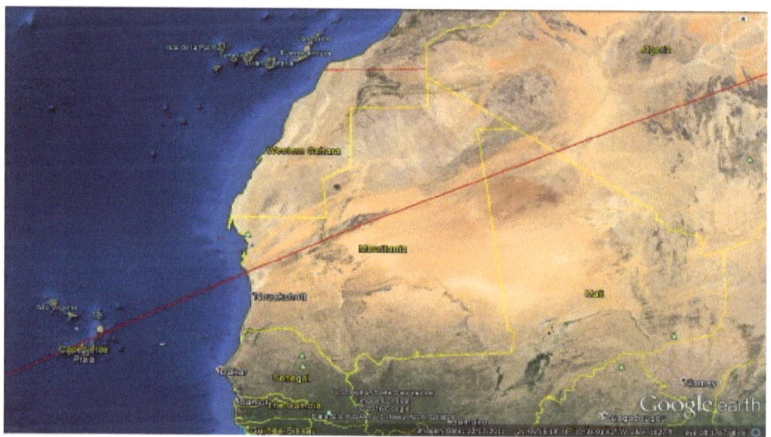

Fig 24

MACHU PICCHU CONNECTS TO SAHARA DESERT

Nobody seems to know why there is so much Sand in the Sahara, or where it came from. Scientists seem unable to cope with Construction on such a large scale, and so, deny it's existence, even if it is Staring at their faces!

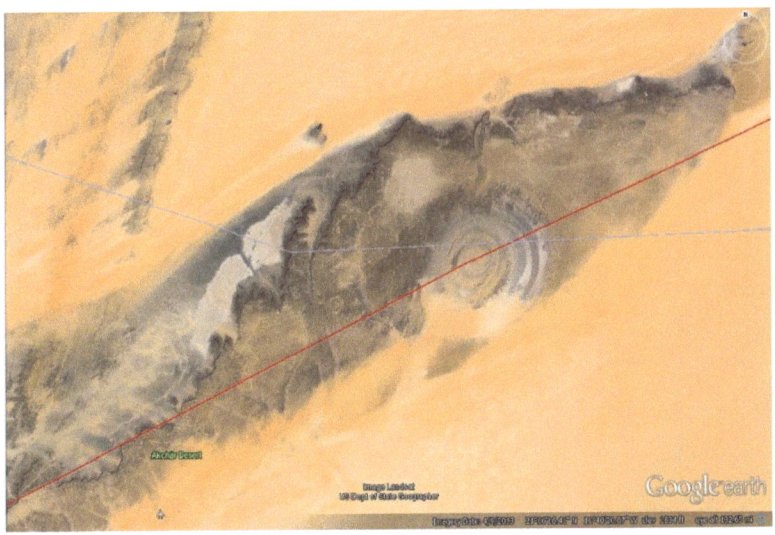

Fig 25

THE GREAT EYE OF THE SAHARA

From the Great Eye of Horus, the path takes you across the Desert, directly over the Pyramid at Giza! This Path is marked, intentionally, and it has higher Meaning and Purpose as we shall see!

This path takes us through the heart Lands from the Koran, from the Bible and Torah, through India a Indus Valley, exiting the Asian Continent via Myanmar, Thailand, Cambodia and Vietnam. Pretty much, this was Historical Path of the Buddha; Siddhartha Gautama.

Fig 26

THE GIZA COMPLEX AND GREAT PYRAMID CONNECT TO MACHU PICCHU

From Vietnam, the Solstice Path goes from one end, to the other end of Papua, New Guinea, where there is, not only Gold, still being dug-up by the Ton, but there are RECENT Stories of Giants, STILL present, on the Island.

One tale has Giants throwing Heavy Equipment, like Children's Toys, and leaving MASSIVE FOOTPRINTS!

But the REAL Proof, of this Solstice Path is the NEXT point along the Journey Around the World; a strange little Island in the Middle of the Pacific Ocean called, Easter Island! Remember, I started at Macchu Picchu!

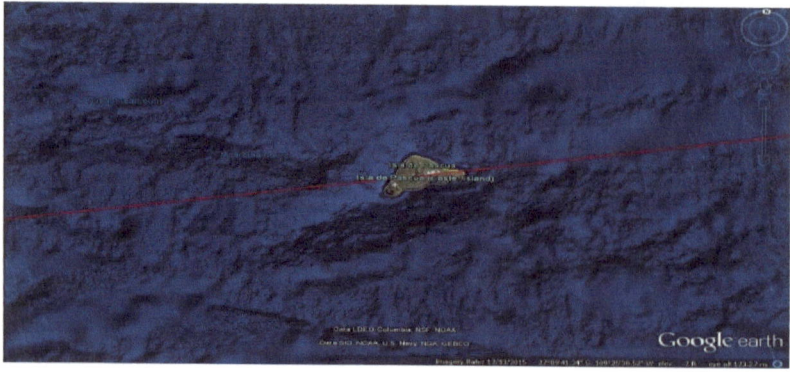

Fig. 27

SAME SOLSTICE LINE CROSSING EASTER ISLAND

From Easter Island, back to Macchu Picchu. At this point, I have to note, the Path comes to the South American Shore near Paracas, and the Ica Desert, of Nazca Lines, fame.

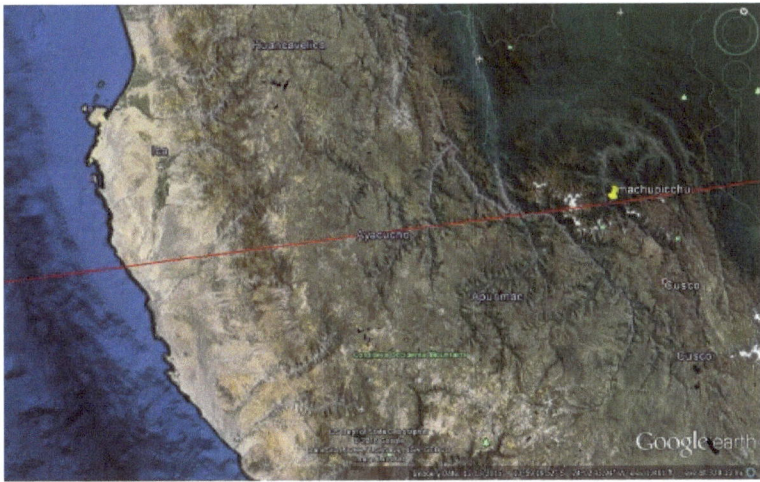

fig 28

SAME SOLSTICE LINE EASTER ISLAND CONNECTING PARACAS DESERT TO AND MACHU PICCHU

This Path, has been noticed as the 'Giza Meridian', and remarked upon by Leaders in Egyptology, and even Brian Forester, of Inca Tours, and Paracas Museum fame, has said, "nobody knows why it Lines-up". I have apparently found the Answer.

6. CONNECTING THE DOTS

In the Emerald Tablets of Thoth, there's a Phrase, that Thoth declares; 'Let Us Make a Religion...", so that none forget. If you've ever Studied any 'Solar' Religion, you've notices they all follow Sun patterns: Equinox and Solstices are just two of any number of examples that can be told. These similarities, and equalites, found in all Sun Worshippers, that are followed by Peoples as Disparate as Indidgenous Peoples, to Nordic Vikings. Peoples who have Historical Meetings, hold the SAME beliefs for ONE reason ALONE: It's an EASY SYSTEM to TEACH! As Simple as Drawing a Circle, and putting a Stick, in the Middle, to show how to follow the Shadows!

The Stories that come FROM those Peoples, tend to say the SAME thing; that gods came from the Sky, and Taught us. I'M not asking anyone to believe without knowledge, the same information available to me, is available to ALL.

Continuing to seek Verification that the Paths I was following, were actually there, although NOT visable, came as I again, went back to Santa Catalina's Stonehenge Layer to see if the other Lines of 'Import', specifically, the Lunar Lines, taken from the Blueprint

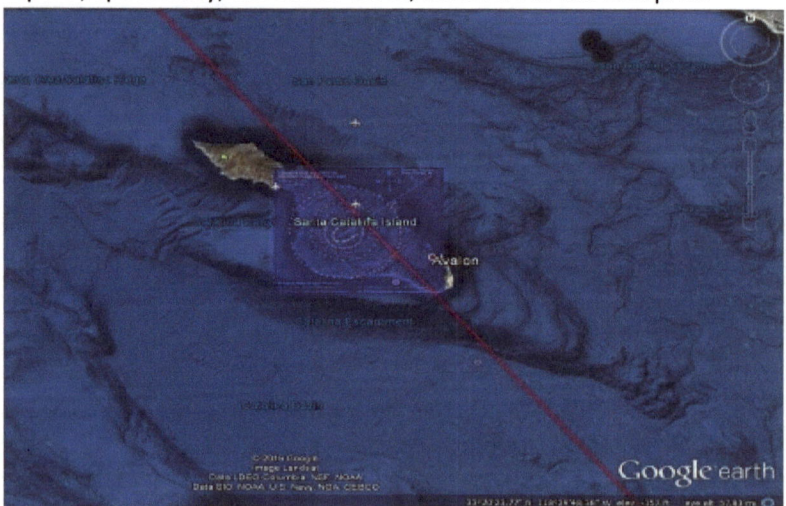

fig 29

MAKING A LUNAR CONNECTION WITH SAINT CATALINA

Fig 30

LUNAR LINE FROM CATALINE CROSSING WELL USED VERTEX AT SANTIAGE, CHILE

The Path from the upper Left of the Image, comes from Santa Catalina, Calif, while the Path, just under, and to the lower-left comes from the Adam's Calendar, South Africa. The importance of these Paths, crossing in Santiago, Chile is; Historically, it is one of the most Ancient City-sites in the World! Santiago, dramatically shows how Man, has used the same Geographical areas, time again, each new group Building, on the Foundations of Older Civilizations. Mayan Walls with no Mortar, lie under repairs done in Inca Block, and 'newer' topping those are, the Brick and Stucco of the Conquering Spaniards, complete with Crosses on top!

Time to go back to Stonehenge, UK, to verify that these other Paths can be followed to anything meaningful, otherwise it could easily be counted-off as, curious coincidence. I tried a North-South Line first just to see if anything meaningful was there, it connected DIRECTLY on a north-South line was the Center OF Spain's Peninsula, exactly the Path coming from Santa Catalina! Too many coincidences!

After seeing that the Island Continent of Madagascar, can be seen as a Giant Footprint, and that the whole North American Continent can be used as Stonehenges, I decided to Layer the Penninsula of Spain. It looked like an exact match.

Fig 31

DRAWING A NORTH/SOUTH LINE DIRECTLY FROM STONEHENGE CROSSING CENTER OF SPAIN

The Path that intersects Spain's Stonehenge is the Path from Santa Catalina. When the Solstice Line is drawn, it aims DIRECTLY for Moscow, Russia. And when extended across the Atlantic Ocean, once again crosses Santiago, Chile, at the SAME point of crossing that the Adam's Calendar Path crosses!

Fig 32

SPAIN AS MASSIVE AS STONEHENGE

The People who 'connected' the Earth set-up Shrines along EACH of these Paths, as I could clearly see that you could easily follow the Paths of Migration along these Lines, too.

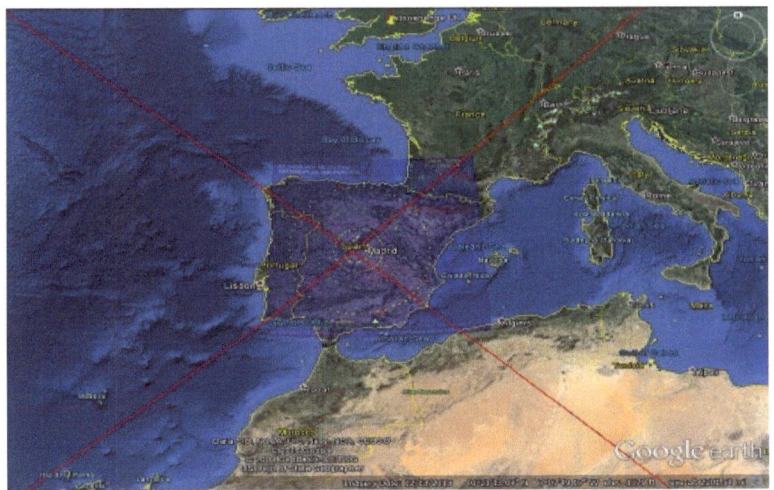

Fig 32

SPAINS SOLSTICE AND EQUINOX PARTS

That Humans had walked these Paths, just like I pictured, came from a Friend, who, while researching her Indigenous Heritage, sent her DNA to be Tested for Genetic Background, here are the Results:

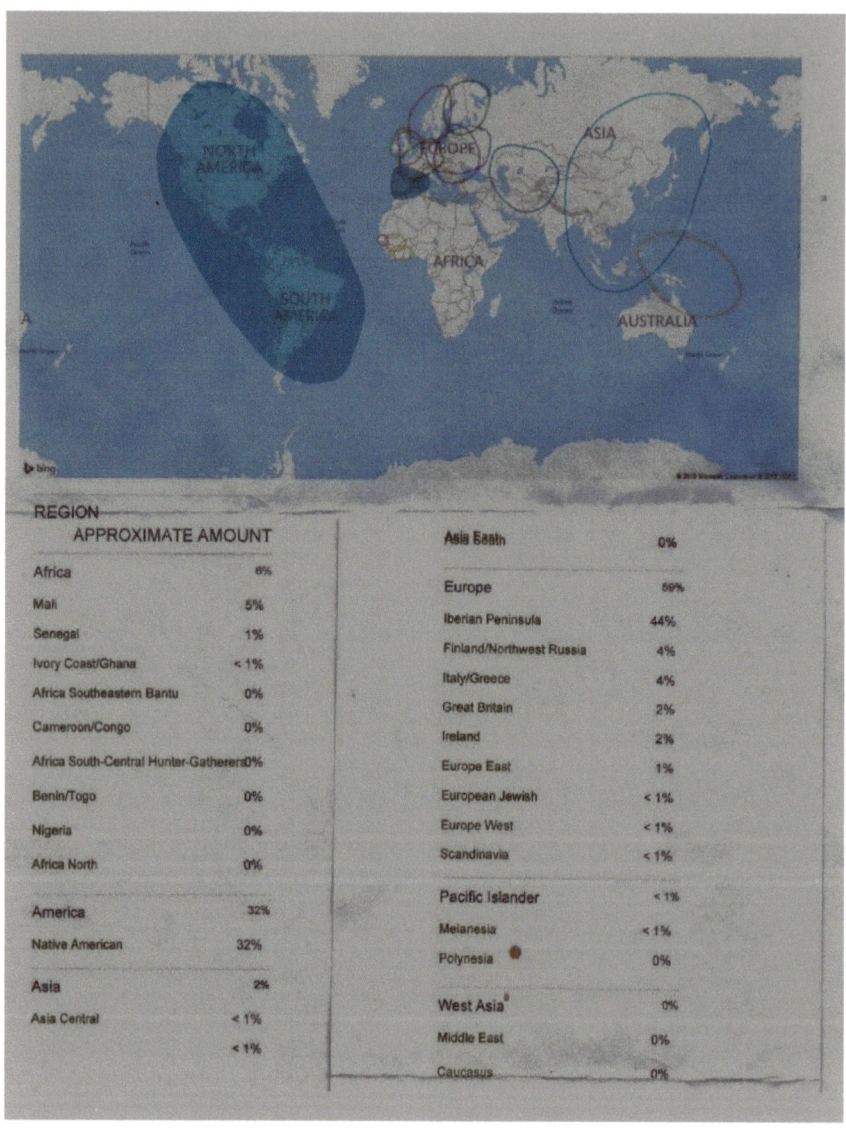

REGION	APPROXIMATE AMOUNT		
Africa	6%	Asia East	0%
Mali	5%	Europe	59%
Senegal	1%	Iberian Peninsula	44%
Ivory Coast/Ghana	< 1%	Finland/Northwest Russia	4%
Africa Southeastern Bantu	0%	Italy/Greece	4%
Cameroon/Congo	0%	Great Britain	2%
Africa South-Central Hunter-Gatherers	0%	Ireland	2%
Benin/Togo	0%	Europe East	1%
Nigeria	0%	European Jewish	< 1%
Africa North	0%	Europe West	< 1%
		Scandinavia	< 1%
America	32%	Pacific Islander	< 1%
Native American	32%	Melanesia	< 1%
Asia	2%	Polynesia	0%
Asia Central	< 1%	West Asia	0%
	< 1%	Middle East	0%
		Caucasus	0%

Fig 34

FRIENDS DNA FOLLOWS SOLSTICE PATHS

I can easily overlay these results, with the Spain, and Santa Catalina Paths. It turns out that her DNA shows both North, and South American Indigenous Peoples; 32% , 44% Iberian Peninsula, and by lower Percentages, following the Adam's Calendar Solstice Path. Using JUST those 3 Paths, and cross-checking Percentage of Genetic material shown, I surmise that; Her People started in the African area (showing the least DNA) moving along Adam's calendar Path. Then to South America, through the Pacific Islands, while the OTHER half of her Heritage travelled from the United Kingdom TO Russia, from the 'lost' Empire of Tartaria, to Greece and Italy, settles in Spain. Western history tells the rest of the Tale: Spaniards come to the Americas, Her DNA says her Ancestors were Apache, or Northern Mexico/Southern USA, so that is where her Spanish Ancestry meets Northern Indigenous, extrapolating from her DNA Percentages (showing the most Genetic Material).

So, now I have a Secondary Validation, from a truly 'Scientific' source. I looked more closely at the areas that these Paths crossed-over. Immediately, I was shocked by the Presence of Polygonal Lines, in the Land, wherever the Solstice Lined were, they looked JUST like the Mayan Wall Construction!

7.   THE FALSE ILLUMINATI LINE

When I first started a validation process for the Solstice Paths, I ran into an artificially 'created' Line. It consists of the Placement of 4 very famous Cities, in the United States, apparently intended to link the 'Old World', to the New land they intended to Colonize. These Cities are Named: Boston, Massachusetts, New York, Washington, District of Columbia, and, Roanoke, Virginia. I have found that this Line is linked to Atlanta, Georgia, also.

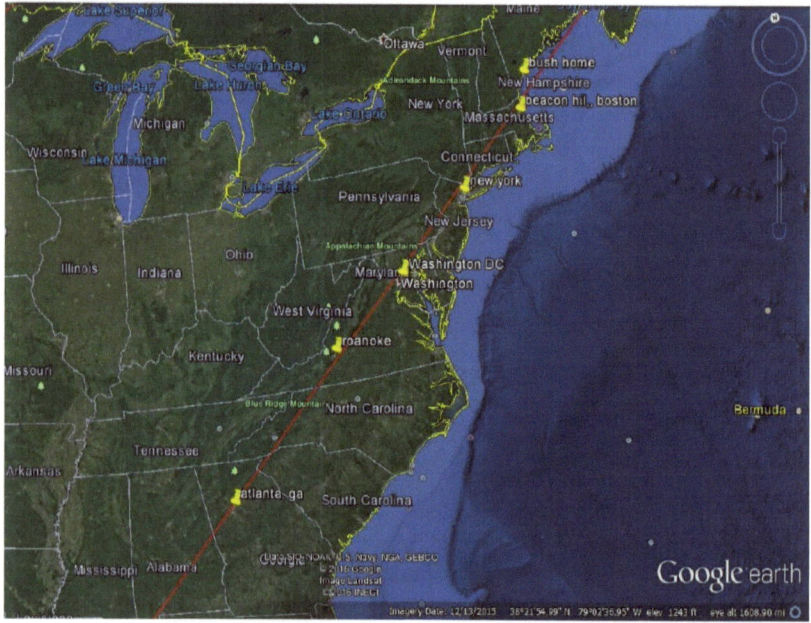

Fig 35

ORIGINAL COLONY SETTLEMENTS HERMETICAL LINE FROM STONEHENGE

Of Note: The Bush Family Home seems, coincidentally, built near where the crossing of the Line takes place over Kennebunkport, Maine.

But, the larger vision of the 'creators' of this Line, was linked to Sir Francis Bacon, and further, to the Mystical Knowledge of Hermetics, as Interpreted by a Particular Man, Named; John Dean. A Man of somewhat devious Influence, in the Courts of Nobles, was also an Interpreter of Esoteric Knowledge coming out of Egypt, and was obviously, someone who interested the Genius of Sir Francis Bacon. History's acknowledged Hero of the colonization of the New World, Sir Francis Bacon was John Dean's foremost Student of Hermetics.

Hermetics can be found in most Tarot Card Systems, and is easily discernable as Egyptian references, although Artistically insinuated. Without trying to explain the System which you can easily find if interested in the Subject, I contend that; Medieval Peoples, lacked the Tools, or the PURPOSE of these Lines. I submit that, the Illuminati, through Templars, and later, the Masons, must have thought that by simply Re-Purposing the Obelisks they ransacked from Egypt, they would, somehow, POWER-UP some kind of Ancient 'Force', to create a 'New' Atlantis!

When the Knights, known as Templars, invaded Jerusalem, they found Treasures, not only a ton of Gold, but Ancient Artifacts, later misinterpreted as 'Magics', and 'Esoteric' Knowledge which they used to further their own purposes. Somehow, they thought they could build, 'New Jerusalem's' by changing the Geometry of certain Cities where their Influence brought them certain 'Power', over others.

Paris, France, is one such example of the Templars trying to build a 'New' Jerusalem another city with these same 'Measures' is London, England. But, the most famous examples MUST be: Washington, DC! ALL appear to have Egyptian Obelisks! One can even be found in Central Park, New York, another In San Francisco, California, where you can also see, very plainly, a West Coast Pyramid, called the Transamerica Building!

If this is ALL just a little TOO 'coincidental' to you…. I'm GLAD I'm not a better Mathematician!

Altough I'm positive, that these 'Illuminati' have attempted to link the Dome of the Rock, In Israel, to Washington, DC, I'm just as positive that they missed! I did a Stonehenge Layer 'center' of the City:

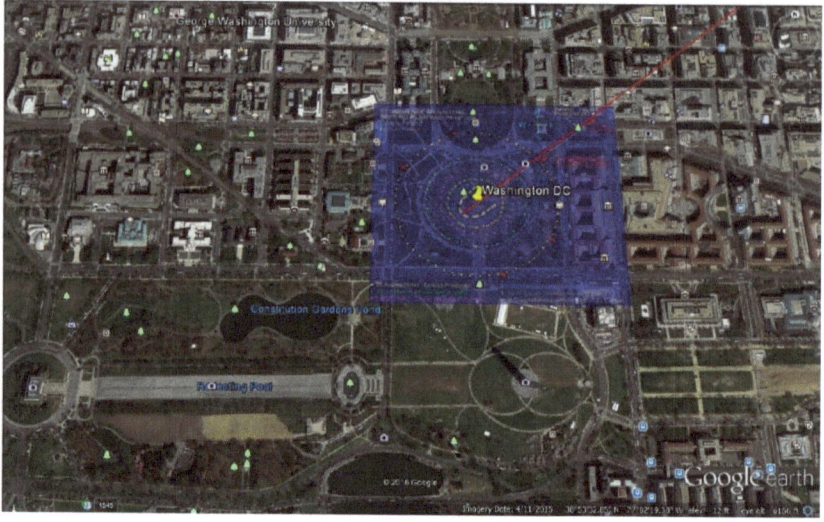

Fig 35

WASHINGTON DC STONEHENGE OVERLAY AT WASHINGTON MONUMENT

Notable is; the Orions' Belt look, it makes, from above, it is also, one of the few places of interest, that Google Earth does NOT 'Kant', or view slightly from side-on.

You can easily see the Washington Monument as from 90dg directly overhead, in the foreground of picture. If you view the Great Pyramid at Giza, the picture from Google Earth is slightly from the side, and therefore, can not be considered as a true view from overhead. AND you now know the 'real' reason for Israel's 'influence' over American Politics.

I followed the Solstice Line (not Path), from this layering, and it DOES cross over Jerusalem, just not at the Dome of the Rock, as I believe they were trying to do. This Line is interesting in that, it crosses over the Grave of Atlas, in Turkey, but also, over a Man-

Made Made Monument to World War 2, that looks, for all the World like a Duplicate of the Georgia 'Guide Stones' USA! The Line ends in the Sinai Peninsula, crossing over nothing of Interest, until, it crosses over an Island, in the Pacific Ocean called; Campbell Island.

ALL, of which, is NOW 'obfuscated', or, NOT viewable, on Google Earth at close-up, I 'Pinned' them, as soon as I found them, so at least you know WHERE they are Located:

Figure 37

HERMETICAL LINE CROSSES GRAVE OF ATLAS

Ancient' Greece is an apparent target, to these 'Hermetics', as this Line crosses directly over Crete, and bisects the Sinai Peninsula. Were they given incorrect meanings, or did they just make-it-up?

Eitherher way, they got it wrong, from the meanings of the Lines, themselves, to their purpose of 'connecting' these sites.

Obviously, the People who were using the True Paths, were not 'Hermetics'...

I won't bother showing Camp Isle, simply because Google Earth that whole area under constant Clouding, even if you DON'T want to view Weather. Further Research into the Templars connection to this Island, can't be accomplished by Satellite Imagery.

On the North American Continent, this Line passes over the Gulf of Mexico, into Mexico, proper, and Mexico City, itself.

Fig 37

LINE FROM WASHINGTON DC BISECTS SINAI PENNINSULA

Of interest is where this Illuminati Line takes you: El Caracol (the Circle) is a huge Spiral feature, and, I believe was the 'Destination' that the Templars must have considered; Atlantis because of it's Circular appearance!

Fig 39

FALSE ILLMUNIATI LINE THROUGH EL CARACOL, MEXICO

They must have been informed of the Structure, from the Spaniards who originally the amazing Mexico City of the Aztecs, Olmecs and Mayans, but, you'll notice that the L does NOT cross the Center! THIS, more than any other Evidence that I have found, sho although the Masons, Templars, Illuminati and Hermetics...were incredibly knowledge they were wrong, even if only slightly!

So, considering these People 'omnipotent', or in-possession of Mystical Magics, is CLEARLY FALSE! More likely, they have come to believe their own nonsense, so, even if don't believe, their muddled-thinking WILL affect you. Any Illuminati 'knowledge' is mo to 'Voodoo', and there are NO 'Zombies', to be frightened of... just 'crazy' people, who only have the 'power' we give them.

8.   WHERE THE EVIDENCE LEADS

By this time I was thoroughly addicted to making Solstice Lines across the Earth, and had Criss-crossed the World so many times, it was starting to look crowded.

I decided I should look into the Polygonal Lines, found wherever the Solstice Paths crossed, when the NewEarth Youtube Channel, among a few others, noted their presence. I note the NewEarth Youtube Channel because they have done great Research into the Ancient Peoples, specifically, from a European view.

 My research 'spanned' the Globe, Polygonal Lines were EVERYWHERE! I was finding evidence that they were 'scorched', in many, if not MOST, areas, and something unusual. Almost ALL the Polygonal Lines had, what appeared to be, small round Water Wells, wherever they connected. Here are several examples:

Fig 40

AUSTRAILIA THESE POLYGONAL LINES COVER BOTH HILL AND
VALLEY. HUMAN CONSTRUCTION ALWAYS GOES AROUND
OBSTACLES.

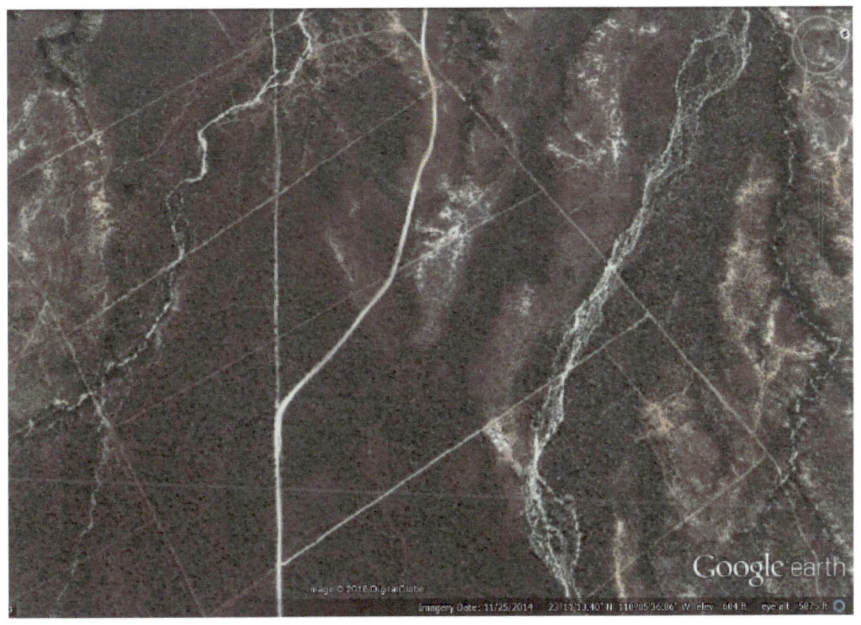

Fig 41

BAJA, CALIFORNIA

Fig 42

MEXICO

Fig 43

SOUTH AMERICA

Fig 44

SOUTH AMERICA

Fig 45

NORTH AMERICA

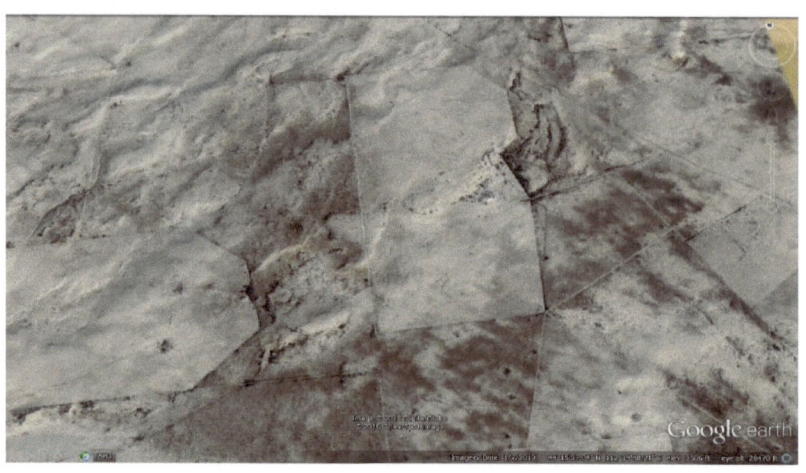

Figure 46

RUSSIA

In Russia, I found some of the most significant signs of destruction, possible AIMED at the Polygonal Lines. I made a video for Youtube called, Nazca Lines Explained; https://youtu.be/qbYFrFFtu5I. In which I detail many more areas of Polygonal Construction, done on a MASSIVE, indeed, MEGALITHIC manner!

Fig 47

POLYGONAL LINES FOUND IN SOUTH AFRICA

This Video also details Conclusions I will get to, later in this tome. Also, I will show more AMAZING things I've found along the Solstice Paths like the following Picture, I call, Layout For A Pyramid.

Fig 48

POLYGONAL LINES HIGHLIGHTED

I had found the Line feature, in Google Earth's toolbox, so I Highlighted this one and this is what it look WITH the Lines all 'lit-up'.

22 Years in Construction, to a Journey Sheetmetal Worker, who spent 4 years in Trade School, learning to turn Sheets of Metal, into usable items, this looked, for all-the-World, like something could be bent, at the Joints, and turned into a Pyramid...albeit a VERY large; with the Lines of the Triangles measuring almost 5 MILES!

BUT, if these Lines were that significant, to the Anunnaki; remember that I had link them to Stonehenge, in the FIRST place! Then I linked the Henge to the Solstice Lines, which linked to the Giants of History, NOT Mythos, and thus to Megaliths, around the World, all linked by these Solstice Paths. I started to closely inspect areas of the Earth, wherever Paths crossed, with Genetic evidence, all things connect to the WAY that these Peoples crossed the Earth!

AND, if ALL things are connected, then the MOST Famous of Megaliths...the Great Pyramid at Giza, should connect, too...

9. THE GREAT PATHS

Those who have Studied the Great Pyramid at Giza, know it contains hidden Mathematics, of an INCREDIBLE Design. Part of that Math is; ALL PYRAMIDS ARE CIRCLES. That's right, Spherical Math IS represented by the Great Pyramid!

What MOST People are unaware of is; ALL North Facing Pyramids have 'built- in' Summer Solstice Lines! So, it only made sense for me to Hi-light them, around the World:

Fig 49

FROM THE GREAT PYRAMID, CENTER OF EARTHS LAND MASSES

The prior Picture is the Giza 'Center Of All Land Mass (fig 49). These Lines are draw EXACTLY through the Apex of the Pyramid, matching Corner-to-Corner, and extending the Globe.

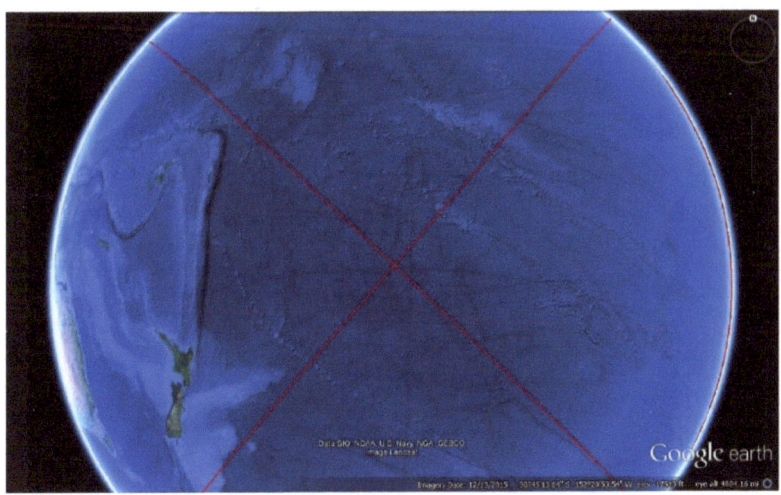

Fig 50

LINES FROM THE APEX TO CORNERS OF GREAT PYRAMID LEAD TO CENTER OF EARTHS WATER MASS

Picture, above, is where these Lines cross, on the Exact Point, 360 degrees from the Pyramid.

 Of Note: this is the EXACT CENTER OF WATER MASS ON EARTH!

The next Picture shows how the Lines, drawn from the Apex of the Pyramid, to the Edge of the Corner, and extended, actually 'CUP', Egypt's Fertile 'Crescent'!

Fig 51

LINES DIRECTLY FROM APEX OF GREAT PYRAMID

The Summer Solstice Plath from the Pyramid takes a short swim across the Mediterranean Sea, aiming DIRECTLY for Haifa, Israel.

Fig 52

SOLSTICE LINE FROM GREAT PYRAMID POINTS TO HAIFA

Across Syria, Southern Turkey and Northern Iran, the Path leads to Asia, through the Caspian Sea. I want to Point-out that these areas are where the United States has been concentrating their Military Forces, and I just HAVE to wonder if anybody in some 'Secret Program', has been aware of this, and simply excluding the Public? It might even help Stop the destruction if the Public were to become aware of their connections, to one-another? This Path crosses Asia, exiting through the Island Empire of Japan:

Fig 53

LINES FROM GIZA CONNECTING TO STONEHENGE AND RUSSIA

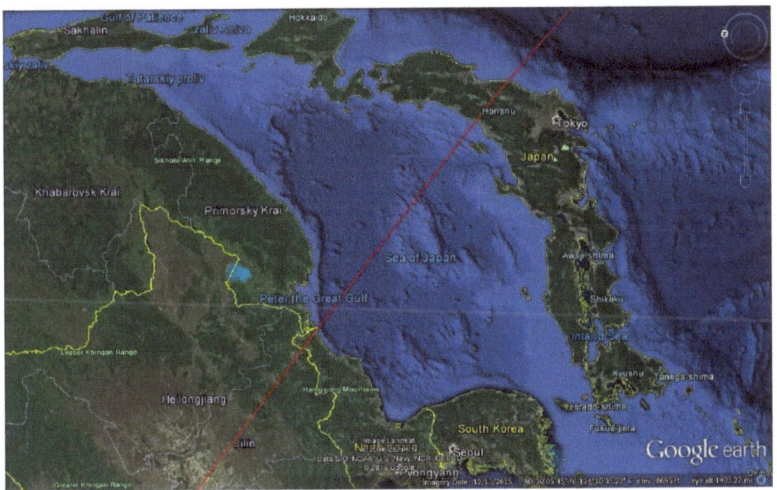

Fig 54

LINES FROM GIZA THROUGH RUSSIA AND HONSHU, JAPAN

Where it crosses the center of all the Earth's Water Mass, and then to South America:

Fig 55

LINE FROM JAPAN CROSSES TIP OF SOUTH AMERICA

Then BACK to the Great Pyramid, through the Heart of Africa

Fig 56

LINE FROM SOUTH AMERICA CONTINUES BACK TO GAZA

BUT, the real interesting Path comes from the 'Winter Equinox' Line, drawn, and extended across the World. THIS is the Path connected, EXACTLY, to the CENTER OF STONEHENGE.

Fig 57

FROM NORTHWEST CORNER OF THE GREAT PYRAMID THROUGH EUROPE

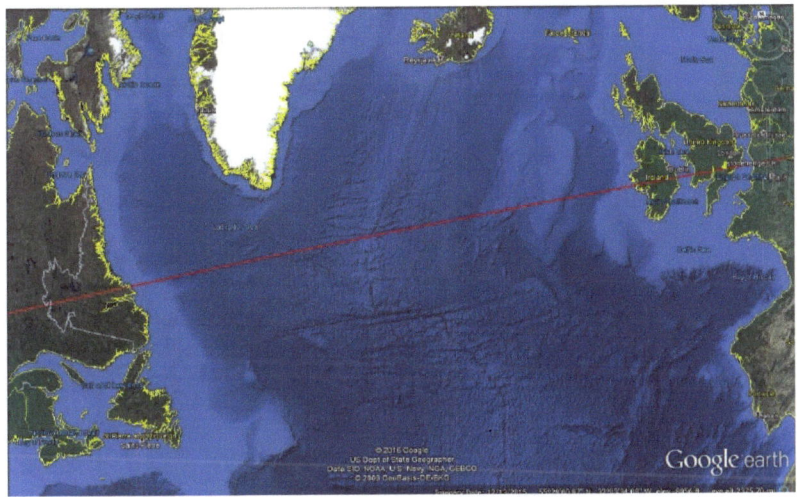

Fig 58

PATH TO NORTH AMERICA DIRECTLY FROM GIZA

This is where I have to think of this Path as a 'Western' Path, or, the Path that was followed, to the 'New Land':

Fig 59

PATH FROM SOUTHEAST CORNER OF GREAT PYRAMID THROUGH
SINAI

At this time, I became acquainted with a Book called; "The Anunnaki
Of Nibiru", by Gerald Clark. I started listening to the many
Interviews Mr. Clark had done on the Capricorn Radio Shows, and
heard him speaking about the laboratory, chronicled by the
Sumerians, in the Sinai Peninsula, the Anunnaki had called, 'EDEN'!

There comes a time when the Term; Coincidence, no longer has
meaning, and literally becomes an 'IM-possibility'. Perhaps you've
heard THIS Phrase, famously uttered by Law Enforcement: ' ONCE, is
a coincidence, TWICE, is a Plan...THRICE...is a CONSPIRACY!'

I had already passed the Point of 'Skepticism', and 'Un-belief', and
was actively checking the Surface of the Earth, wherever these
Solstice paths crossed, but after the shock of Aden, I was following a
Path of Ocean, until, finally, the Path crossed Baja, California.

I want to leave this to the Reader, to Validate, because, where this
Path enters the Land, are more Polygonal Line Construction, but an
apparent, UNDISCOVERED, Pyramid!

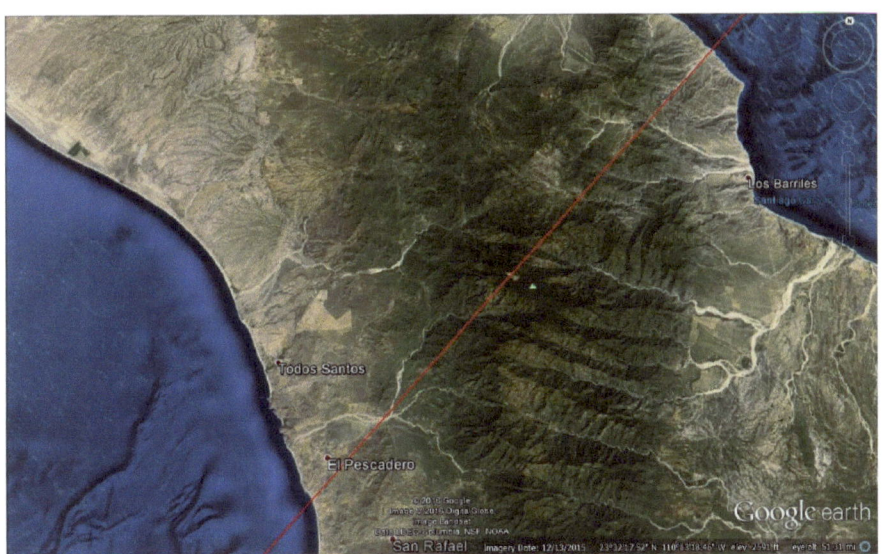

Fig. 60

PASSING THROUGH SOUTH BAJA FROM GIZA

Matter-of-Fact...there are LOTS of Pyramidal Structures, just WAITING (hopefully, Credentialed People will make the effort, before Thieves)! Through the Sea of Cortez, the Path crosses leads to Central Mexico.

Again, Polygonal Lines, at the Coast, where the Path crosses, lead me on, where I 'pinned' a location, I called a "bitchin pyramid". Here, I made another Stonehenge blue Layer, to verify I was, indeed, following a 'planned' Path, and to connect the Path, with Path into New Foundland, Can, where it leads to Stonehenge, proper, and to the Great Pyramid.

Fig 61

GIZA PATH BAJA INTO MEXICO

Fig 62

FULL GIZA PATH THROUGH NORTH AMERICA

I made the 'Find' of a Lifetime, when I found the Polygonal Lines, both IN, AND ALL AROUND the area of the Chihuahua Desert, Big Bend National Park and Presidio County, Texas, ALSO known as...

10. MARFA TEXAS

Marfa, Texas, was first established in 1883, as a Water station, also the first documented sighting of the phenomena known as the 'Marfa Lights', and which the area is now most famous for. The Marfa Plateau, just Southeast of the City, proper, is at an Altitude of 4,685ft, or almost exactly 1 1/2 mile above Sea Level. Population in the 1930s was approx 3900 people, and in 2010, the US Census reported approx 7,800, just about doubled in about 100yrs. Still, sparsely Populated, by any standards, and "Minimally Developed", by their own admission. By comparison, the Nazca Plains, of Peru, are only 1,710ft, or, just under 1 mile above Sea Level, but the yearly Rainfall is about equal, on the Marfa Plain.

The only 'real' development, with the exception of the addition of the Railroad, occurred in the 1940's, due to World War 2, when the Marfa Airport was built, just to the East of Town, and is now, the location of the Marfa Lights Viewing area. Only to be abandoned, just a couple of years later, had a short stint, known as the Alta Vista Airport, which became important, as I will later detail.

It has been determined that the City was named after a Character in a Jules Verne called; 'Michael Strogoff' (unfortunately, this Author hasn't read), the Character was a Family member Named: Marfa Strogoff.

A Cattleman named Robert Reed Ellison, and his Cowhands, were Camped in the area known as; Mitchell Flats, 1883, first reported seeing the Lights, thought them Apache Campfires, but upon Investigation, found nothing to indicate a reason for the Lights. Although, this would be the first, of many, who've chased the Lights through the years, literally, when a couple of Airmen had reported a Night Flight over the Plateau. They were apparently trying to track the Lights, but had to abandon the chase, after a couple of close-calls of crashing the Plane.

There has also been some 'serious' Research, that has, also, failed to come-up with answers; these range from Archaeological Students determining ALL the Lights were Car lights, without answering why they had been reported BEFORE people, historically BROUGHT IN

Artificial Lighting, to the area. Some Geologists claim that the Lights are a 'natural' product of the Terrain, this area has been famous for it's Volcanic Scenery, also had a bit of Silver Mining, until early 1900s. There have been very famous People involved in very expensive Documentaries, including a segment on CBS's '60 Minutes' Show.

But the BEST evidence of the Marfa Lights, come from Edson C Hendricks, an Electrical Engineer with a Masters' Degree. He went to Marfa, in 1991, with the intent to Research the Phenomena, even to the Point of building a Device, for Detecting Electromagnetics. I have been in direct Email contact, and he has graciously allowed me to use his Research, which I re-print, in total, as he mailed to me:

Fig 65

Seeing the Marfa Lights on February 2, 1991

Edson C. Hendricks

San Diego, CA 92109-2357, USA

marfa@edh.net

December 26, 1991

On Saturday, February 2, 1991, I visited the Marfa Lights viewing area east of Marfa on Route 90. I brought with me a good quality pair of 9x35 binoculars and a simple homebuilt receiver for audio frequency electromagnetic radiation, similar to radio signals but at much lower frequencies. My reason for having with me such a receiver was that atmospheric electrical phenomena are known to operate at such frequencies in some cases, and I wondered whether there might be a connection between such radiation and the Marfa Lights.

I arrived at the Marfa Lights viewing area, parking at the far right at about 5:30 in the afternoon, while the sun was still low in the sky. At this time the weather was clear and dry, with a dark cloud bank low on the distant west horizon. My intent in arriving this early was to gain familiarity with local land features in daylight so that I could more easily interpret the positions of distant lights after dark. Using the binoculars occasionally as darkness gradually descended, and ranch lights and automobile headlights switched on, I quickly learned to identify each by their position and appearance. The ranch lights were still and constant in brightness, and were either white or blue in color. Most of the visible automobile headlights were from northbound traffic along Route 67, 20 to 25 miles to the southwest. These moved at an apparently steady rate in a northward direction, following a constant path that gradually descended toward the north, and were white in color. These lights appeared to vary in brightness, evidently due to roadside obstructions and veering angles along the distant curving road. Along this road there were several points at which headlights often seemed to flare up brightly, no doubt as the road path directed the headlights toward my position. I tracked several automobiles as they followed the distant road, and soon I could accurately anticipate how the apparent headlight brightness would change as the automobiles proceeded.

Seated in my parked automobile, using the binoculars, I continued to examine the landscape in this manner until daylight had almost completely faded. At several minutes before 7:00 P.M., I suddenly noticed a rather bright light to the west- southwest, north of the region where the automobile headlights were visible. I had seen no light in the same vicinity prior to that moment. The light attracted my attention mostly because of its color which was brilliant yellow, unlike the other lights I'd been examining. I carefully situated the binoculars on the headrest of the passenger seat next to me, so that they were quite stable and directed toward this new light. Although it was late dusk, I was still able to see a nearby utility pole in the same field, providing me a fixed reference point with which I could gauge the light's movement.

Initially the light did not appear to move, but there did seem to be a slight erratic variation in its brightness. As I stared at it through the binoculars it began to move slowly to the north. It remained brilliant, perhaps somewhat brighter than the brightest headlight flashes I'd seen along Route 67, but it continued to flicker slightly and irregularly. The light gradually seemed to gain speed somewhat, then it gradually slowed as it passed behind the nearby utility pole in front of me. As it emerged from behind the pole it slowed to a halt, and remained stationary for half a minute or so. Then it slowly began to rise and move back toward the south, back behind the utility pole and past it to the south. It had by now ceased rising as it moved laterally southward, still flickering but it was now generally gaining brightness. I was puzzled by this light, but I had not yet been able to decide whether I might be seeing another automobile headlight. And then, as I stared with great surprise, the light divided into two separate lights which continued to move southward and gradually drew apart.

At this point I glanced at my watch, noting the time at 7:08, so I had watched this light for about ten minutes. As I continued to watch what was now a pair of lights, the leftmost one (to the south) flickered a bit, quickly increased brightness, and divided again. At the same moment, the rightmost light of the original pair (to the north) abruptly changed direction, began to move away to the north accelerating to a speed much greater than any automobile headlights I'd observed earlier, and gradually faded in brightness. It vanished about as it reached a point due west of my position, almost in lin with Route 90. By then as I looked back toward the other pair to the south, yet another similar light had appeared, and the erratic movement and varying brightness continued.

I watched what seemed to be a fairly continuous display of this light behavior for another twenty to thirty minutes. I recall seeing as many as five lights at one time during this phase. I noted that the central point of the display seemed to be moving gradually toward the south. During the entire performance I could observe slowly moving automobile headlights in the distance which were distinctly different in appearance from the lights to the north that held most of my attention. Gradually the lights faded away, and by about 7:35

it seemed to me that only ranch lights and a few distant headlights remained.

As these peculiar lights were fading, several other automobiles arrived at the viewing area, parking to my left. So I got out of my car and strolled over to greet the new arrivals, who were also standing outside their cars. I asked if any of them seen the lights, and was startled to hear that they were still visible. I looked around to see that indeed the performance had resumed, but perhaps even more extravagantly than before. There now seemed to be generally more lights visible at once, and these had moved further southward toward the distant headlight track along Route 67. Moving headlights were visible along with these anomalous lights, but were easily distinguished due to their color, motion and brightness variations. Use of the binoculars made the distinction much easier to perceive.

The others present were evidently seeing just what I saw. There were two who seemed familiar with the phenomenon, whom I assumed to be local residents, and several others who seemed to be tourists. There was an open discussion among the group of what each person was seeing; all reported seeing exactly the same light behavior, and all reports agreed with what I saw through the binoculars. I then returned to my automobile to continue observing the performance, which continued for perhaps another twenty to thirty minutes. Finally these lights again faded, at about 8:00 P.M. I remained and watched carefully for further appearances. During this period I was still able to see all the ranch lights and distant automobile headlights, but I saw nothing else that resembled those lights that moved erratically and glowed with the often brilliant, flickering yellow light. During this hour the other visitors evidently lost interest and gradually departed, as did I at around 9:15 P.M. I noticed rain appearing on my windshield even before I arrived at Marfa ten miles to the west, and the rain continued intermittently for most of the night.

During the hour between 7:00 and 8:00 P.M. I observed a display of evidently inexplicable moving lights that was nearly continuous and fairly complex from time to time. At some points the scene rather resembled a three-ring circus, as there was more concurrent activity in different directions than one could fully follow. I recall at least six instances of seeing a single light divide into two that gradually moved apart. In one of the instances the two moved quite some distance apart, perhaps six degrees from my viewing position, and then both reversed their motion and moved back together. As they met, one on the right appeared to spiral upward abruptly, seeming to circle above the one on the left, flickered out and vanished. I observed several instances in which a light vanished abruptly, and another appeared abruptly at a distance of several degrees laterally from where the light vanished, seemingly at the same moment. If the light actually moved the distance, it did so at a speed so great that I could detect no trace of any motion. In one case I noticed a barely visible, dim deep red light appearing to move southward from one light toward another adjacent to it. These two were each a brilliant yellow, and I believe the emerged when an individual light divided. As the faint red light approached the vicinity of the leftmost bright light, it flickered and suddenly flared up, changing color to the same bright yellow as the other two. It moved past the leftmost light and proceeded gradually southward.

At most times during the hour-long display multiple lights were visible. The individual lights would sometimes hang motionless for short periods but most of the time they would move. Such movement was generally erratic, involving both gradual and abrupt changes in speed and direction. The lights appeared to move both horizontally and vertically, but individual lights always seemed to move over much greater distances horizontally than vertically. So far as I could determine, this motion exhibited no directional preference. Motions of individual lights often seemed independent of other lights, but there was an apparent tendency to form straight lines of equally spaced individual lights which would often seem to move together as a group. At one point I counted as many as six separate lights visible in such a line for a short time. I noted cases of different lights moving apparently independently that would abruptly move into a straight line formation, and cases of individual lights departing such formations and diverging off independently. In one startling instance, an entire row of four or five bright lights disappeared from view at the same instant, evidently extinguished as though controlled by a single switch. And then, after only several seconds, evidently these same lights began switch back on, but this time individually in no apparent order or pattern.

None of these lights ever seemed to be very close to my viewing position. I doubt that they were closer than several hundred yards, nor further away than perhaps fifteen miles. The lights were not beyond the hills in the distance, as at times appeared in the foreground against them. Nor can I make any reliable estimate of the physical dimensions or altitude, except to observe that they never appeared to rise very far above the horizon along the distant hilltops.

Some lights glowed steadily for fairly long durations, but they usually would flicker and brightness's would vary irregularly and erratically. I did note one thing that may have been a significant pattern. Prior to the moment that I observed an individual light to divide into a pair, in each case I recall, the original light rapidly grew very brilliant. I began to anticipate seeing a light divide when I noticed this quick brightening, and it usually did so.

The binoculars I used displayed a 7.3 degree field of view. At a distance of twenty miles, for example, the width of the field would be about two to three miles. At most times the multiple light displays could be observed entirely within this view, but in a few cases individual lights were spaced at extremes beyond this limit. I noticed that the central point of the display of multiple lights seemed to be quite stable. This point was initially west-southwest from my viewing position, but gradually drifted to the south that it was southwest by the time the display finally faded an hour later.

In my judgment I was watching neither ranch lights nor automobile headlights, and others present unanimously agreed. I spent about three hours watching the same vicinity.

During the first and third hours I saw many lights, all of which appeared to be ordinary ranch lights and headlights. These same lights were fully visible during the second hour, but something strikingly different from these was also Virtually all of the automobile headlights I saw appeared to move at slow, near-constant speeds in a northward direction only. These other lights moved erratically, often shifting speed and direction. The patterns of movement I saw were inconsistent with any reasonable explanation involving automobile headlights, and the colors of the two were quite distinct to my eye. Of course the ranch lights were very easy to identify, since their position, intensities and colors were absolutely constant.

It is conceivable that the appearances were due to atmospheric mirage effects, but this also extremely difficult to accept. The colors of the lights that moved in peculiar patterns generally tended to be hues of yellow, unlike any other apparent light source.

I do not immediately understand how atmospheric image distortion could involve consistent and stable color changes. Perhaps a more convincing argument against atmospheric distortion effects can be made in the case where the light display had moved directly to the southwest from my viewing position. That is the same direction as the distant Route 67. At the time the peculiar lights appeared interposed between my eye and that distant scene, which included frequent automobile headlights and a fixed tower with a steady red light beneath a flashing red light. All were perfectly visible at the same time occasionally the images would appear to pass quite close to one another. Using binoculars I examined these very carefully, and I detected absolutely no trace of distortion or attenuation of any distant images. I can understand no way that atmospheric mirage or distortion effects could generate such peculiar light images, while concurrently transmitting adjacent images of objects in the distance showing no perceptible distortion at all. I observed the usual desert mirage effects occasionally in the vicinity, but none during the time I spent at the Marfa Lights viewing area, and none that seemed to be associated with the peculiar light behavior in any way that I could discern.

I was listening to the audio frequency electromagnetic radiation detector (or "VLF receiver", meaning "Very Low Frequency") for the entire duration I spent watching for these strange Marfa Lights. It seemed to be working properly, as it easily detected emanations from the nearby electric power lines, as well as VLF noise from some of the passing vehicles. I had hoped to find some fairly strong and constant signal that might point to a power source for the peculiar light phenomenon, but I heard nothing such as that. However, I did notice the frequent occurrence of rather loud "whistlers", which are signals caused by a lightning strike (sounding rather like a bullet ricocheting off a rock) weather was clear at this time, but a cold front was approaching from the west and nearby.

Beyond the horizon to the northwest, I noticed occasional bright flashes I interpreted be due to lightning strikes. I also heard crackling typical of lightning noise on a standard AM-band radio I brought with me. It seemed to me that the whistler signals were surprisingly strong, considering the apparent distance of the storm, but I'm not sure this is either relevant or true. However, it did appear that the occurrence of such signals evidently due to the distant electrical storm activity correlated with the appearance of the peculiar lights. That is the two emerged roughly concurrently, persisted over approximately the same period abated at about the same time. I am certainly not convinced, but I would not yet rule o possibility of some causal connection between the distant storm activity and these strange lights.

In closing, I am a professional electrical engineer, computer system designer, and scientist. As such, I am always skeptical toward "unexplained" phenomena. However, I am absolutely certain that what I herein report having observed was nothing that can be readily explained. My main reason for visiting Marfa was to investigate the reports I'd read, hoping to determine if these might easily be dismissed as hoaxes or errors. I determined that they could not be, but I surely did not expect to see the phenomenon myself with so little difficulty, and in such a spectacular display. I now wonder if I was extremely fortunate, or if the Marfa Lights are perhaps less elusive than one might expect such an extraordinary phenomenon to be. I plan to return in the near future to pursue this and the many other questions.

An Addendum -- Whistlers and the Marfa Lights

I wrote the above report immediately upon my return to San Diego from Marfa on February 3, 1991, while my memory of the circumstances and my perceptions was very fresh. Since then I have edited this report only to clarify its language and to correct grammar and spelling. I have taken care to avoid modifying its content in any way, and I stand by account's completeness and accuracy to what I believe I encountered in Marfa that evening.

Some months have now passed, and during this time I have been investigating a v of issues related to this peculiar matter. For the most part, nothing I have learned since February 1991 inclines me to qualify anything I reported my above account. However, regarding the topic of whistlers, I now believe my earlier account conveys an honestly erroneous impression. I shall not modify the content of my original report, although it be in error, but instead I shall append here my clarification.

The above report includes my observation of the "frequent occurrence of rather loud `whistlers,' ...." I paid little attention to these sounds because I did not recognize them abnormal, as explained below. As I now recall, these whistler signals occurred perhaps once every five seconds or so averaged over time, and their typical duration was between half second and one second. I recall their timing pattern to have seemed random and irregular.

At the time of my February 1991 visit to Marfa, my understanding of whistlers and other atmospherics was only cursory. I understood whistlers to be signals generated by lightning strikes which generally follow geomagnetic field lines, and which can cause peculiar audible effects that sometimes resemble the sound of a bullet ricocheting. I had never before attempted to capture any whistler signals, and I had no real awareness of their properties. When I heard what I believe I identified correctly in Marfa as whistler signals, I merely assumed they arose from the electrical storm activity perhaps 20-50 miles more distant. I noticed the distant lightning, heard signals, presumed the two were connected concluded all was normal.

It turns out that all was not normal. I had designed and built my very simple VLF receiver on the spur of the moment, without notable research or analysis. My intent was not to find whistlers, but rather to seek the possible presence of some strong, probably constant audio-frequency electromagnetic signal that might plausibly power atmospheric discharge effects. I had never attempted to ascertain anything regarding the signal sensitivity of this device prior to my trip to Marfa; I relied solely on my belief that anything capable of driving a bright discharge effect, ought also do something audible to that device. I am still quite certain that belief is valid.

When the device began to produce very loud whistler sounds, I did not realize the signals must have traveled a very long distance beyond the atmosphere, nor that they must have been arriving from the southern hemisphere. A whistler's characteristic concurrent decay in both frequency and amplitude which persists a second or so is caused by its traversal of perhaps hundreds of thousands of miles along geomagnetic field lines lying well beyond the earth's surface and atmosphere. So, my simple presumption that these signals were arising directly from the distant, visible electrical storm activity was definitely incorrect. The local storm may have been generating them if they were propagating to the southern hemisphere, reflecting and propagating back to Marfa, which is surely possible but it is by no means certain.

In fact, I was suspicious of these signals even as I heard them. They seemed to me too strong to make sense of my assumption that they were being driven by lightning activity beyond the horizon from my location. This reservation does appear in my original report. I noticed, but did not report, that I could not correlate audible whistler signals with any visible lighting flashes. At that time I was fully engrossed observing the Marfa Lights. I confess I overlooked the lack of detailed correlation between my observations of the lightning a whistler signals, and that my tacit assumption of some connection between the two was without firm basis. In fact, I have no good reason to connect the distant storm activity noticed in Marfa with either the signals I heard from my VLF receiver or with the peculiar lights I watched at the time, but I still cannot rule out the possibility of some kind of relationship.

My initial suspicions led me to test my VLF receiver's sensitivity using some professional laboratory equipment immediately upon my return to San Diego from Marfa. I was initially surprised at the device's very low sensitivity, since the whistler sounds it produced in Marfa seemed loud. Still, the measurements were clear; this device produces loud sounds only if the electromagnetic audio frequency signals in its vicinity are quite strong.

I observed these anomalous phenomena at the site along Route 90 designated for purpose by the Texas Historical Commission. Directly adjacent to this site runs an aerial power transmission line of moderate scale. Judging by its appearance, I would expect it to carry substantial current, and if so it would accordingly radiate a very strong 60 Hz. signal throughout the vicinity. This VLF receiver easily detected the power line's field, but the apparent whistler sounds it produced were much louder, easily drowning out that 60 Hz. noise. In regard this in retrospect as highly peculiar, and I can offer no explanation.

I have subsequently experimented with this very same device in the presence of electrical storm activity in San Diego. Since my return from Marfa early this year, at least five significant electrical storms have passed through the San Diego area. I have been careful to listen to the output of the VLF receiver I used in Marfa on each of these occasions, and heard nothing resembling the sounds it yielded in Marfa. Each of these storms passed closer to my location than did the electrical storm I observed at a long distance in Marfa. One of these storms was extremely intense electrically, and it passed directly over my San location. This one caused a power outage that began just as the storm approached major intensity in my vicinity, and the outage lasted for several hours. Many powerful lightning strikes very close to my location were obvious from their flash and noise effects. At the same time, since power was out, my VLF receiver was reporting almost no extraneous signals. Still, the only signals it produced were some rather faint "click" sounds concurrent with nearby lightning hits. I heard nothing from it resembling a whistler.

I am unsure how to interpret all this. I don't really understand how whistler signals could power persistent, long duration, brilliant atmospheric discharge phenomena, and I do not conclude that they do. On the other hand, I don't yet understand why such evidently anomalously strong whistler signals would appear in Marfa just as the odd light phenomenon emerges, then fade just as the light phenomenon fades, with no further explanation. After I heard these signals I actively attempted to detect them again until I left Marfa late the following day, with no success at all. I regularly monitor this VLF receiver now at my home in San Diego, but it has produced no whistler sound since the evening of February 2, 1991, in Marfa. I still suspect some association between the whistler signals and the Marfa Lights I encountered concurrently; but I no longer presume that either involved the local electrical storm activity at the time, nor can I dismiss the possibility that they might have.

REPRINTED BY PERMISSION, USE OF MATERIAL DOES NOT IMPLY ENDORSEMENT

Although he wasn't able to identify the Phenomena, Mr Hendricks WAS able to car detail an event, NOT possible for "Car lights". Mr Hendricks now resides in San Diego a California, is disabled and retired. He is still, a delightful, highly intelligent person. After researching the Marfa Lights, this Research stands as a Scientific Validation of the Phenomena.

I didn't find the Marfa Lights, I can't hope to match this amazing Research that Mr Hendricks provided, but WHAT I found might EXPLAIN them...maybe explain...everything.

One more item of Note: Marfa Lights cannot be seen under a full-Moon, so if you're going out to view the lights, use of a Calendar would be helpful

11. A NORTH AMERICAN 'NAZCA'

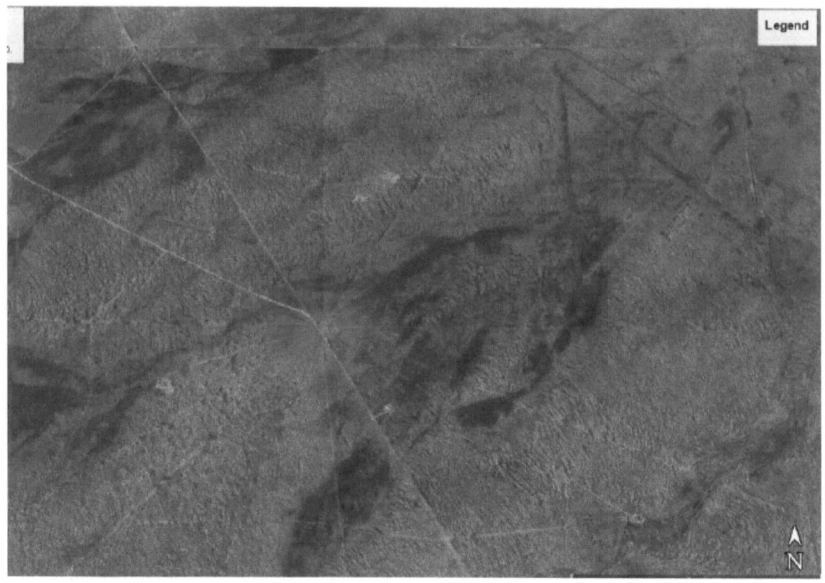

Fig 64

UN-RETOUCHED EASILY VISABLE LINES ABOVE THE MARFA PLATEAU

The very first thing I saw, from Google Earth, was the same kind of Polygonal Lines construction, that I had found in other areas around the World. To view them better, I used Google's 'Line' tool to highlight them. The area of dark Lines, in the upper-left of Picture, above, was, obviously, to me, as a representation of a CROSS!

Fig 65

CROSS AREA

At this time, I must mention that there are people Residing here. Because Texas Laws of Privacy, any area of obvious Habitation will only be depicted via Google Earth, but I believe that these are Archaeological Sites, that need further Investigation. I believe that all the Lines I found are ART, intended to be viewed from Altitude!

JUST like Nazca, Peru, these Lines only make sense when you see them at various from Ground Level. And, like Nazca, when Lines are viewed from the Ground, they blend into surrounding Geography. BUT, unlike Nazca, these seem to use TEXTURE, Geography, AND Lines, all to make HUGE 'Geoglyphs', of Animals, AND 'Faces'. The Construction of this Art includes massive areas of different 'color' of Soil. And NONE of it has been Tested!

As a very apt Metaphor: imagine being Ant 'sized' person, with a limited vision, never seeing that you happen to be standing in the Mona Lisa. Not only, would you be unable to comprehend what the Picture meant, but the Person depicted, wouldn't be recognizable as a person, unless the Ant-person were, somehow, to attain a proper Altitude, without going too High, or too Low.

This area of Southwest Texas, as mentioned previously, is a Site of varied Volcanic activity, and the Terlingua area is Notable for the many Fossils strewn about. I have come to believe, after trying to contact Archaeologists, Texas State University Archaeology and Anthropology Dept., that the Geology, must really be an impediment, to their Researching this area, as they don't want to be the First to find Fossils linked to these Lines!

If Fossils are above or below the Geophysical Art, then it will be a determining Factor of Ageing, or dating this massive Construction. Can you imagine having to be the Archaeologist who has to inform the Public of extreme Antiquity, related to the Discovery of possible Technology...still working?

According to United States Census Bureau, this is an area of 'Minimal Improvement', and after trying to find Paperwork relating to Road Construction, have determined that NO major Blasting, or Bulldozing, ever had to be done because the Ground is almost like some kind of Concrete, all Construction was basically a re-paving.

This area, has, quite a few 'ranch Airports', which, seen from Altitude, indicate more Ancient usage than they now enjoy. When seen from Altitude, the 'Runways' seem part of a much larger System, often as their Geometry actually matches, kind of a figure '4', Pilot will recognize as an 'Airport Pattern'. Some of these 'Airports', are found underwater, as the picture, of the Sea of Cortez, shows;

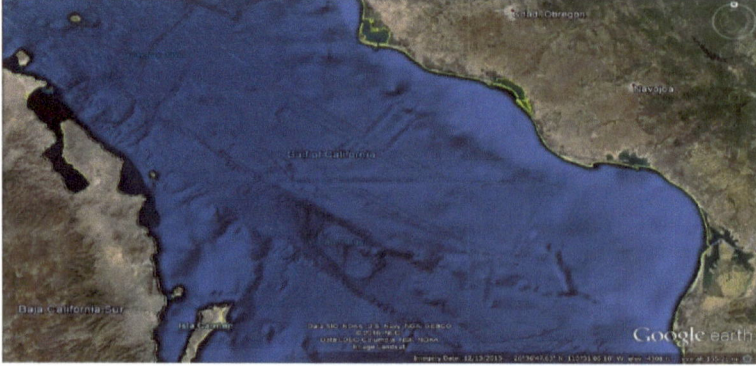

Fig 66

APPARENT LINES IN SEA OF CORTEZ

In Effect, this area reflects the same Pattern, Human Beings follow, wherever they settle; we Re-use the existing Construction, for the Present use. Places all over the World, such as Santiago, Chile, which shows Ancient Megalithic Walls, then Aztec Construction, then Spanish and, finally, 'Modern' repairs. Basically, any area of any 'use', will be Re-used over and over if built to 'last'.

The Link to the Video I made about the Runways and 'Airports', is;

https://youtu.be/SMHN1GiHjEl

The Term: Megalith, means any Structure, or Construction, of Great age, made Stone. Nazca, Peru, is considered 'Megalithic', also because of the size of the area of construction. I have found that, the Marfa, Texas area, meets or exceeds ALL of the Conditions of 'Megalithic' Construction, and is linked directly to the 'builders', of these Megaliths, throughout the World. Indeed, returning to the Great Pyramid at Giza, remember that, an exactly straight Line, Drawn from the Southeast Corner, through the exact Center of the Apex, extended across the Northwest Corner, travels through Europe, into the British Isles and the exact Center of the Altar at Stonehenge itself!

Following this Path, lead me directly to the Marfa Plateau! As a 'fan' of the Sherlock Holmes Books, by Sir Arthur Conan Doyle, I, like the Great Detective, himself, 'don't believe in Coincidences'. Sherlock would also continually intone the cliche' of 'Occams Razor', which goes; (paraphrased) 'That the Simplest Explanation of an event is usually the Correct Explanation, in which, not One, but all Findings point to the same conclusion: that ALL places are linked by PURPOSE, not by 'coincidence'!

I have previously presented Evidence of the 'Validity' of these Paths, so, now it led the Question of WHY this huge area, was never found?

Was this just an example of seeing things from an 'Ant's' perspective? As a 'true' Skeptic just had to know.

When I first started to Research, I found huge 'Solar' patterns, or Lines radiating from a central Point, and the 'Cross' I had found. These, I thought of as Religious symbols, so seemed easy to see that the Marfa Viewing Area, was built over an older Construction, Google Earth had it named (correctly) as; the Alta Vista Airport.

Anyone could see the Alta Vista Ranch Airport, had been defunct for many years, had gone through a Name-change, to become the Marfa Lights Viewing Area. When I checked with the Federal Aviation Administration, I found that they were maintaining a Class E Restricted Airspace, and the reason for that was to 'protect' a 'Radar Tower', that never existed!

Of course, all I needed was a 'mystery', to make the Hair on my Neck rise! I filed my FOIA (Freedom of Information Act), to locate Documents relating to the Construction of the Alta Vista Ranch Airport. The Response was to Deny they had any Information, but, apparently someone had located Documents relating to the Runways; their Identification Numbers, and Composition. It told me there were Three Runways; two made of Dirt, one of Asphalt.

When I went back to Google Earth, someone had changed the Location, of said, defunct Airport. I had found the reason for the Class E Airspace too!

The FAA, or someone IN POWER appeared to be keeping this area hidden!

So, I did the 'natural' thing and; wrote to the Regional Head of the FAA, in the area, Randolph P Loveless questioning if Class E Airspace was still in effect, and also notifying him of the Discovery of the Lines and Geoglyphs, which would probably increase the Air Traffic of Presidio County.

Let me explain what a 'Class E Airspace' is; this is a Restriction against flying in-between the Altitude from 700ft (feet), and 14,500ft, above Ground Level, and can be changed to 18,000ft, anytime, without reason. What this means to a pilot is; when you Take-off, you immediately, attain an Altitude of Fourteen thousand feet, as fast as possible!

Remember My Metaphor about the Ant? In order to properly view this Art, Google indicated an Altitude of 1,000ft-5,000ft was necessary for proper viewing.

I became a 'Champion of Liberty', filing multiple FOIAs, and uploading Videos, at least once per Week, to Youtube. Once, making the mistake of claiming to be a Journalist, the Workers in charge of investigating for your FOIAs, seemed to Deny any Requests, apparent to me their dislike of Journalists. And, Part of the Land they were Restricting from view, was Land around the Big Bend National Park! Land that was being maintained at Public Expense!

My focus of attention was the Alta Vista Airport, specifically, the history of the Construction. I had found traces of older Construction, indeed, the 'new' work appeared designed to use the existing development, so as not to fully obliterate it! And, from Altitude with Google Earth, becomes something like the Jewish Symbol of a 'Star Of David'!

I wanted to know the answer as to the Why, they decided to build in this place? Was it because this area was already Level and brush free? Or, just maybe, someone in our Military saw what I saw, and made a Decision that a Jewish Symbol, in plain view, during World might become a Target for Nazi Missiles? Possibly a little too close to Area 51 and White Sands Military Bases?

But, in this Age of Cell-Phone Technology, ANYBODY can be a 'Journalist', that definition used to Protections under the US Constitution, now, it seems to make Government un- reachable. I started to include the Names of FAA Personnel, in my Youtube Videos, at least One per week, all through September 2015.

Finally, I received a Two-Sentence E-Mail, from the FAA, Thanking Me for informing them about this 'wasteful' Restriction, and, on October 15, 2 the FAA has REMOVED the Class Airspace Restriction!

Fig 67

STAR OF DAVID -AKA- MARFA LIGHTS VIEWING AREA

NOW, these Lines, Art and Geoglyphs are AVAILABLE for Public Viewing, as a True American Megalith!

12. GEOGLYPHS

The term, Geoglyph, means a large design made of Rocks, Stone Fragments, Tree and other durable, long lasting Element. I would add; that a True Geoglyph is only appreciable, from Altitude.

The Lines I was Highlighting, throughout the Plateau, and the, seemingly, Religious Symbols I had found, pointed to a massive 'collective' effort. Other apparent Symbols were: Yin Yang, Tao circle, and a Moslem Crescent, were all found in the same localized area.

To Date I have found and High lit, more than 50 Geoglyphs, since I discovered them in July 2015, as I write, Today is March 7, 2016, but the last Geoglyph I worked-on was in January. I became very busy, when I decided that I had to go Physically, to take my very own Pictures and 'Boots-On' Research, for Validation.

In this Chapter I will Show these Geoglyphs, as I Researched. Here, is the List of this Art:

IMAGES FOUND AT MARFA TO DATE:

ANIMALS

Camel - Goat - Giraffe -

Hummingbird - Seal - Sphinx –

Horse - Buffalo - Porpoise -

Bighorn Sheep & Mate –

Nessie - Bird facing South-

Crested Seabird (Seagull) -Owl

Alligator – Cockatoo

IMAGES

Matriarch- African King-

Priest- Gnome- NIMROD-

Full-Figure Face in Profile-

Wired Face With Hat-

Face With Turban- Female Face

THINGS

Stars- Wired Cross Tao Symbol

Star of David- Moslem Crescent

Boat- Hebrew Nameplate Wired Cutting Tool

Mosaic Cutting Tool- Ankh

Arc of Covenant - Radius Lines

Pyramid Book- Hexagonal Crater

Name/Date Plate- Eye of Horus

Red Squares- Green Squares- Blue Squares

MIXED ANNU/SIMIAN/HUMAN

Fully wired Faces with Annu/Human under 1 Hat

Annu/Orangutan/Human (bearded)

Annu/Baboon/Human

Annu with Solar Symbol on Chest, dark

Simian/Human

Annu Spreading Seed

Lg Annu Mix

Although this List only represents a portion, of the whole, and, what I see, may not be the same, as others will, no doubt, try to Deny that there are ANY Geoglyphs, I will attempt to give the Evidence:

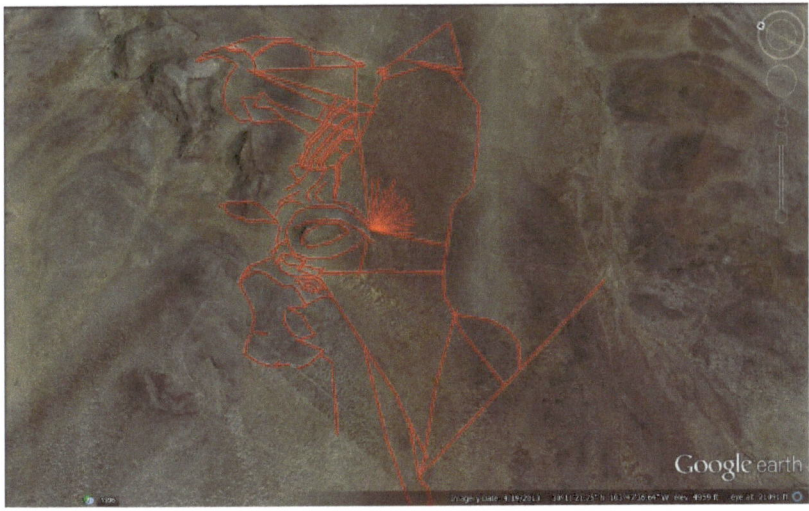

Fig 68

THE MATRIARCH

Fig 69

CONTRAST AND COLOR MATRIARC

One of the most Stunning Geoglyphs, is a feature I call; the Matriarch. I confess to having limited 'Artistic' Ability, but I see a Female, Pharonic Figure. Her Beauty stands more than 2 miles high, and approx. 13/4 mile wide. You can locate this Feature, by using the Co- ordinates given, on lower Right-side of Picture (and all the Pictures shown), just under the Google earth Icon.

The Lines, I high lit, are too regular, too 'Artistic', for any so-called 'natural process' to accomplish. These show-up fairly well, when Photographic enhancements are done, Contrast and Color Saturation, so you can see the Soil Color differs, like a tile, was placed. BUT, this Feature is ORIENTED, EXACTLY, as if the Solstice Path from the Great Pyramid, were both NORTH AND SOUTH! Evidence I followed from attempting Lines from a North/South orientation from the Great Pyramid, had shown only a direct connection South Africa, but no Shrines, or important Megalithic Construction.

The People who built the Great Pyramid, and apparently, all the Megalithic Construction showed no special interest in North/ South arrangements! Egyptologists and the whole (seemingly) 'Lecture Circuit' Experts on Ancient Aliens, had always mentioned the build particular placement, of Megalithic Construction, on a "True North", basis.

Literally, everything I've found, has been because of following the sun These 'Harvesters of the Sun', both followed the Paths-of-the-Sun, but taught Humans to Revere the Sun and the fact, that these many Religious Symbols, were intended to bring together these many differing Peoples, but to Teach Inclusion!

After 22 years in Construction, if there is one, common thing Modern Builders and architects do; is to Orient Buildings on aa North/South Axis! Almost all Human Construction has been, historically, this basis. Most of the Geoglyphs I will show, has no, North/South orientation, and I won't re-orient anything unless, like the Matriarch Image, it must be in that manner. I won't make Evidence, where none exist.

This next Picture appears to be a massive Mosaic made of soil, Artistically Integrated of three Animals; a Two-humped Camel, Giraffe and Nubian Goat. Altogether, this Multi-Glyph measures over ten miles, in Height and Width, and the City of Marfa is built in it!

Fig 69

GIRAFFE/CAMEL/GOAT MIXED ANIMAL GLYPH

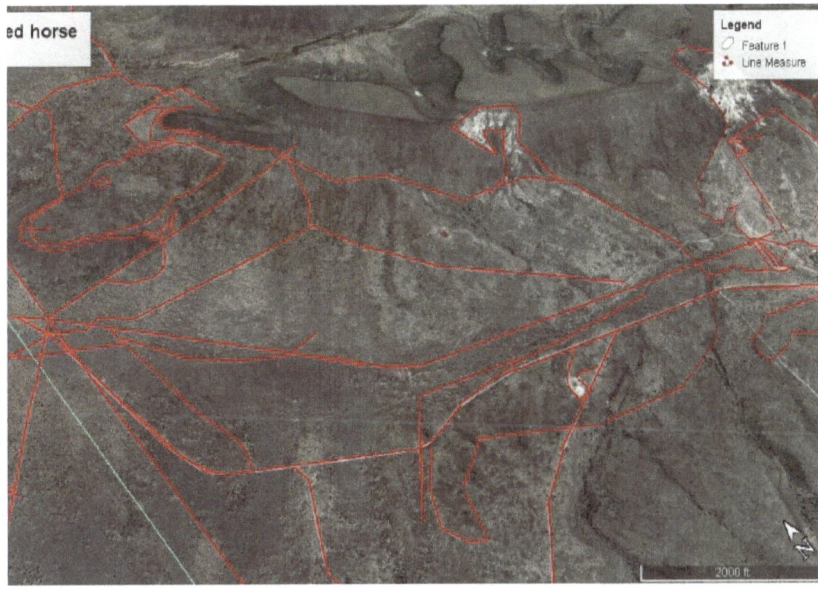

Fig 70

HORSE

Although well- known in Asia, Scientists believe that Camels went Extinct in North America, thousands of years ago. The Two- humped Camel or, Bactrian, variety, is hard-to-find, even in Asia. Fort Davis, was the Historical Home of the Buffalo Soldiers, did, apparently, use some Camels, but remember, the Presidio area around Marfa, had not even have the Waterhole, designation, until the 1880s. And, the Fort was closed for the duration of the Civil War, so, at no time, did Human people, have the chance, choice or ability, to create these Geoglyphs.

This next set of Glyphs should be fairly easy to recognize:

Fig 71

BUFFALO

Fig 72

PORPOISE

The Formations shown are a mixture of Lines, Texture and the Cliff-edge of the Plateau, reminding me of the Artificially formed level Plateaus of Nazca, Peru.

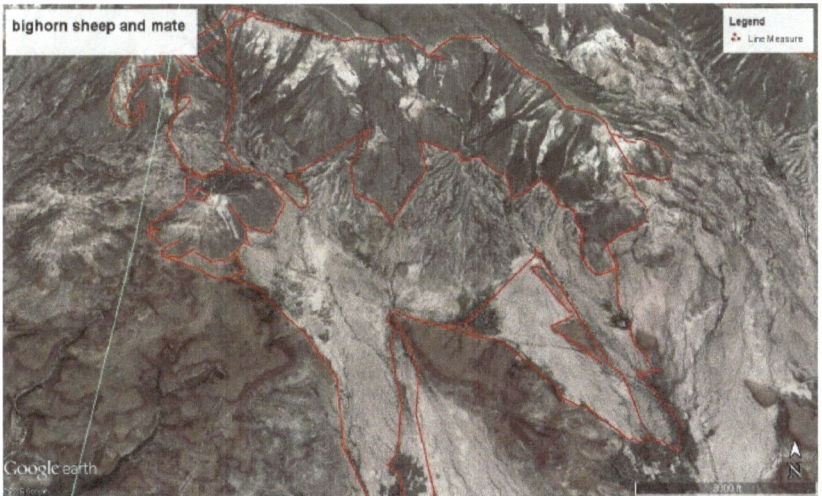

Fig 73

BIGHORN SHEEP AND MATE

I would note that, the 'Mate' under the main, apparent Male form, could not be see Color and Contrast were enhanced! When I did those two Photographic changes, I also that the People, this Art was made for, must have Vision, slightly different, than Human Vision; They must see something of the Full Spectrum of Color!

After, I was able to see, many hidden Geoglyphs, but they will probably be the one Denialists (not true Skeptics), will call "Natural" features.

These last four Pictures are in the Cliff-Face, at the edge of the Marfa Plateau, and as in Nazca, Peru, the Plateau seems artificially altered! Specifically, you can see the tip of the Porpoise's Nose is angled five different angles, to get the proper effect from the Air!

In fact, when your perspective is moved from the South to the North, the Porpoise Dorsal fin, rises, as you move up the Figure (pretty Neat)! It also looks like spray is coming from it's 'blowhole', very Animated, for a Geological 'feature'!

The amount of Material precludes my wish to show ALL of the Geoglyphs, but I will give several examples, of each particular kind of Glyph.

13. MAKING THE ELECTRICAL CONNECTION

" ....for Those who can see with the EYE Of HORUS...All the Mysteries will be Opened..."

(Author Unknown)

This Chapter depicts, what believe, the darker Lines that a found on the Plateau. These are easily seen from the Air, and your Smart Phones, too! As I've shown quite a few people who've expressed disbelief, only to turn to wonder, after seeing the regularity, that only makes sense, when viewed from Altitude.

When I started to Research Geoglyphs, I, very naturally, studies the ones at Nazca, Peru.

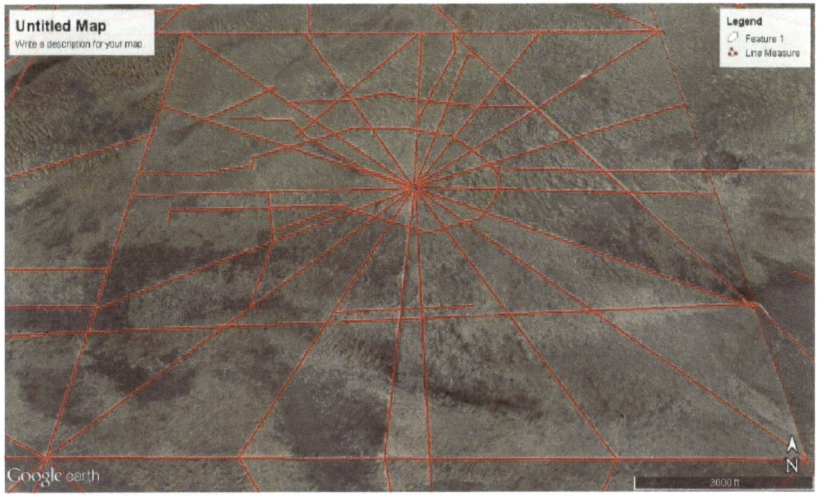

Fig 74

EYE OF HORUS WITH HIGHLIGHTED LINES

Although the Similarities were almost exacting, the Lines I had found, in Texas, were in much better shape! But, there was evidence of obvious Age, and Antiquity, as literally, everywhere, were signs of Erosion. Some of which seemed to be repairs, or work done secondary Date of construction.

I also found the darker Lines were all connected to a small, round, water well. This exact, pattern, was followed, in all of the Polygonal Lines, I had studied, around the World! AND, the Lines of Nazca also had...a Canal System, which I've found, most people are completely unaware of.

This is how Water was directed along the Geoglyphs and Lines of Nazca. Another very famous aspect of Nazca is; Nazca Quartz! You might ask what this has to do with Electricity, but Computer Experts will tell you, it's this particular Quartz, that is being researched for its Storage, Conductivity and Capacitor abilities!

So now, I had some Clues, that the people built these two sites, related only by their similarities were electrical, in nature. The real evidence came from the Chinese.

If you will recall, I mentioned that I had found Polygonal Lines, all across the Earth?

Fig 75

Wikipedia By Ab5602 (talk) Own Work

Use of Material Does Not Imply Endorsement

I had Photographic Evidence, the Chinese had succeeded in Powering-up THEIR Ancient Lines with Hydroelectric means!

Fig 76

APPARENT WATER PLANT, GOBI DESERT, CHINA

Located in the Gobi Desert, this Water Plant is way out-of-place! From this point, it appears that they are forcing a water flow, into a Canal system, which they've un- covered, that run alongside the Lines.

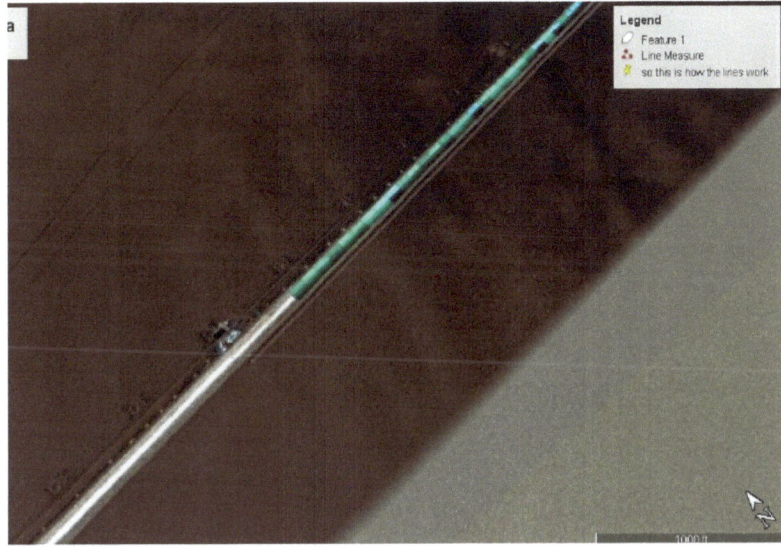

Fig 77

WATER IN CANAL SEEMS TO BECOME VISIBLE BRIGHTER

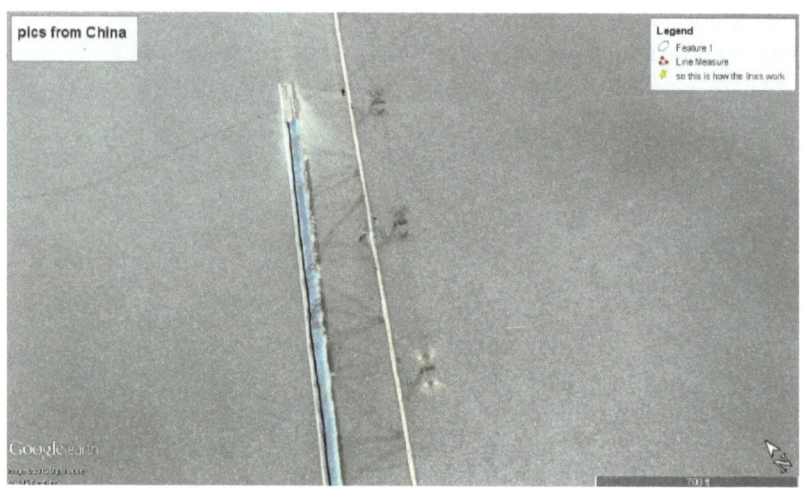

Fig 78

WATER UB CANAL SEEMS TO BECOME VISABLY BRIGHTER

It appears that they have caused some kind of Plasma, or visible effect of Electricity. You might notice that the Canals seem connected, to the energized Line. The co- ordinates for the Chinese Water Plant are: 40*18'39.92 N by, 90*49'51.38 E. At least, until Google removes it!

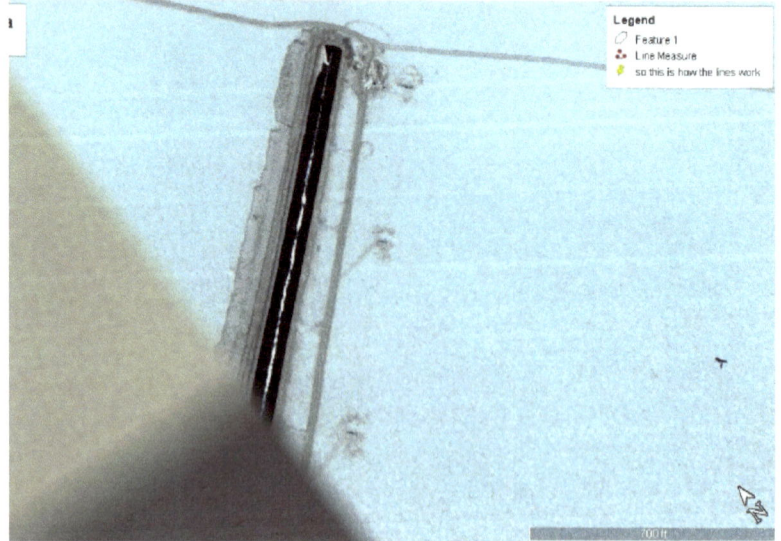

Fig 79

ALTHOUGH NO WATER IS VISABLE PLASMA DISCHARGE CONTINUES

When this is seen, as a whole, at higher Altitude, you see apparent streaking, of the Soil under, and surrounding the Lines. Which leads me to conclude, that the electricity is bleeding off, in the Aquifer, itself, somehow causing it to Light-up!

I now had evidence that the Polygonal lines, around the World, were part of a World-Wide Power Grid! Thanks to the Intellect in China!

I had enough evidence to conclude that: anyplace where these Lines occurred, electricity was plentiful. And, where these Lines were massed, Plasma lighting was expelled, not freely, but controlled!

I base this, as a Statement of Fact, on what I had found in Nazca, Peru, which look like a bank of connectors...a Circuit Board, if you will, and the discovery of, what could only be, Timers, in Marfa, TX! I will picture several, but the sheer number of Inner-Connectors, are just too many. Possibly another Book, this one is about discovery and conclusions, which I get to, later.

I've named them, if you will forgive the audacity. Almost none of these connective structures are North/South oriented, and I have not changed the orientation to make them into something they are not. Some, if not most, seem to be a representational Art, some are actually pictured in the larger Glyphs, as if important enough to picture twice! Once, in small form, again in large.

I have not re-oriented any of these connective structures, all stills are North/South oriented. You will find that almost none, of these connectors are based on that orientation. A fact that is undeniably, not normal, or even Human construction, and my 22 years in the construction Trades attests to that.

Also, you will not see many repetitive connectors, and all seem to have an Artistic Identity! Here we go:

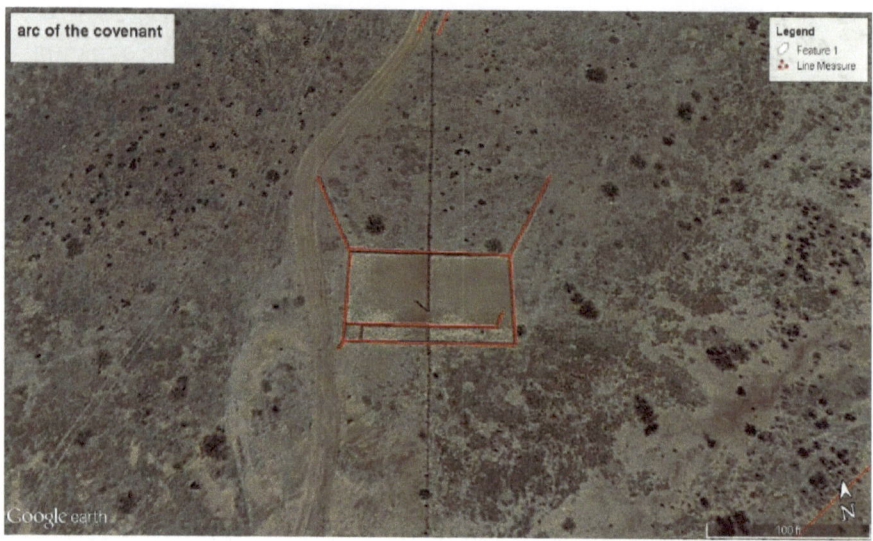

Fig 80

(had to call this one) ARC OF THE COVENANT

Although you can see I titled Fig 81 as "cross" this is actually a close-up of the Radial Image, in Fig 82, on the next page.

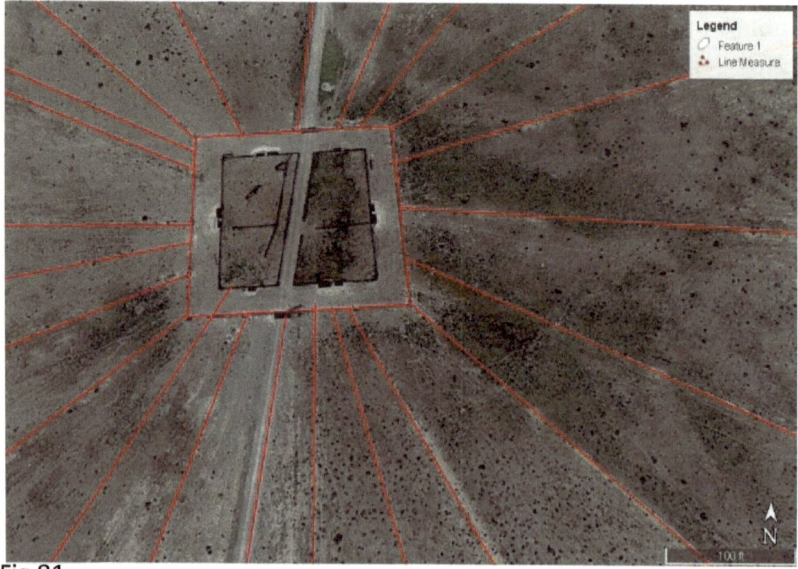

Fig 81

CENTER OF RADIUS LINES CONNECTING PYRAMID BOOK AREA

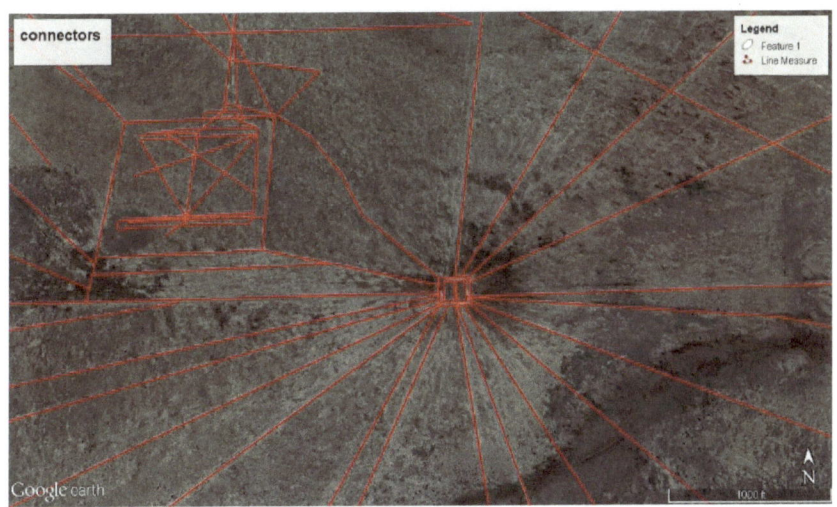

Fig 82

PYRAMID BOOK AREA with a smaller book (Fig 81) as a sun

I have considered this is, possibly, a depiction of the Emerald Tablets of Thoth.

Fig 83

ANGEL, POINTING THE WAY

Figure is found at the 'Top' of the mast of the boat area, where the 'Crow's Nest' was usually located, on Sailing Ships.

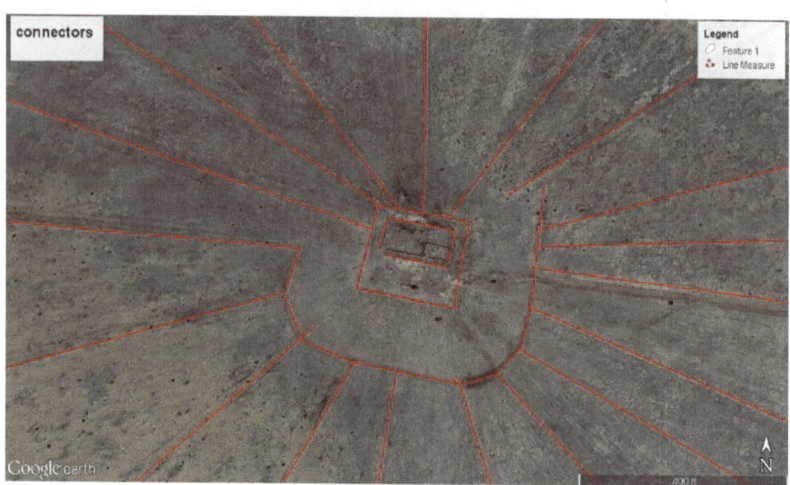

Fig 84

SUN CONNECTOR

There appear to be seven of these major Solar Connectors.

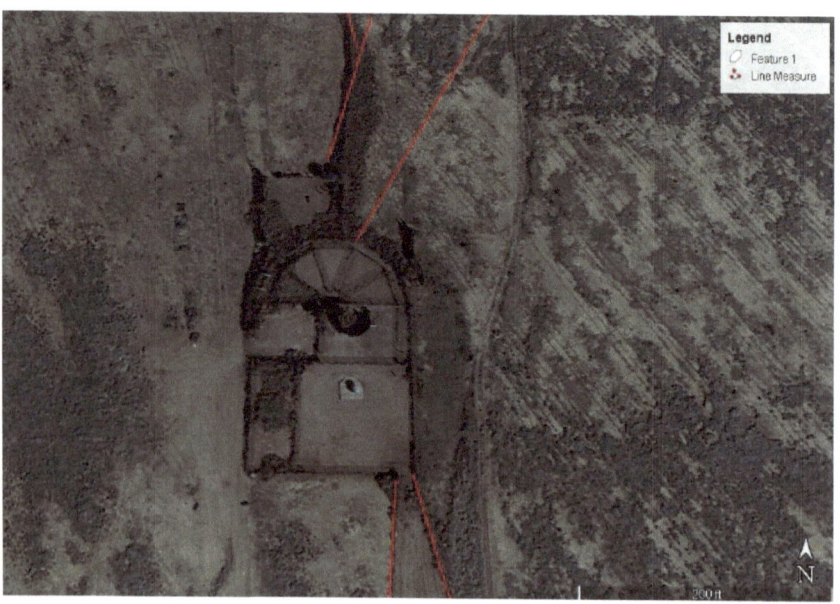

Fig 85

???

This particular sort of face, figure, I have not names, or try to highlight. If you imagine an Electric Current, of sufficient strength, being applied, you can imagine a Plasma extrusion, after a time to build a charge! As a construction worker of 22 years, I can say, with some certainty, that this is not a structure meant for Habitation, or Cattle. Just as Mr Hendrick detailed, so well, the images at Marfa, Texas, and Nazca, Peru, were meant to be an Electric LIGHT SHOW...to be seen FROM ALTITUDE!

Just Imagine the next picture all 'Lit-Up', I believe it shows Who, this area is Dedicated to;

Fig 84

COLOR AND CONTRAST ENHANCED

I see a Wined Sphinx, something on it's Back, and a Seal, all there are Symbols of: THOTH, THE ALANTEAN.

14. TERRAFORMING TEXAS

The Plateau of Marfa has an average Rainfall of approx 12 inches per year. Quite a more than the Plateau of Nazca, Peru, where it averages approx 1 3/4 inch per year. The majority of Rain, in the Presidio County area, takes place in late summer, as a monsoon type. This kind of Rain, caused the area's Residents to invest in a Dam. After much Rese could only find a single instance of Dams, being installed, that Dam is the Saint Esteban close to the US/Mex Border.

Fig 85

SAINT ESTEBAN DAM

BUT, when Satellite images are viewed, ESTEBAN of the DAM area, from Fort Davis, to Alpine, to the Mexican Border, and beyond, there appears to be HUNDREDS, of 'Dam-like' structures. These appear on 'Lines', as well as in the expected Watershed areas, like Arroyos. All seem to be made of a Metallic substance, and, when these Dams, cross Arroyos, almost always are connected to the Lines and Geoglyphs I've found.

Fig 86

DAM OR RESERVOIR ALMOST ½ MILE LONG

Fig 87

CONNECTIVE STAPLE OR SUTURE ARROYO AND LINE

NOTE: this Dam is just North of the City of Marfa, and appears to be outlined, as well as, connected to other Lines. Although no Generating machinery is found, like a human-made Dam would have, it's easy to see that the entire structure is designed to Capture and Control Energy related to Water. (Fig 86)

(Fig 87) This close you can easily see that these lines are NOT used for Roads, nor does it even have signs of ever being a 'road'.

Fig 88

UNKNOWN CONNECTIVE MATERIAL ARTFULLY INSTALLED

(fig 88) HERE, you can see the way Water seems to be used as an Energy source, all part of a larger Geoglyph, and used decoratively!

Fig 89

SUTURES HOLDING OPPOSITE SIDES OF ARROYOS TOGETHER

(fig 89) These Metal-looking structures which cross Arroyos, don't appear to be collecting, or storing Water, like the ones previously shown, but seem to have another purpose: like the SUTURES a Medical Doctor uses to hold Flesh together, when more permanence is needed than 'Stitches' provide.

Fig 90

TWO DIFFERENT SIZES OF SUTURES

There are 3 sizes of these Sutures; from a 1/10th Mile, 1/4 and 1/2 Mile long, subjectively, as I can only trust the Tools given from Google earth. I have had no reason to doubt their measurements, as Google Earth has been known to be used in U S Courts of Law.

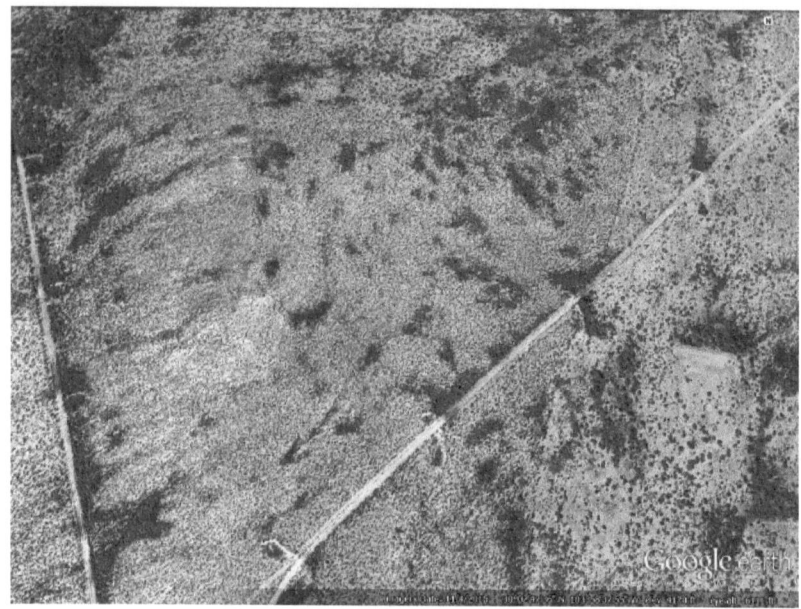

Fig 91

SUTURES, STAPLES AND LINES THAT HAVE NEVER BEEN ROADS

Some of the Sutures show indications that they are working their way out, and actually look just like Staples, not quite driven solidly.

One ne can only imagine the Machinery needed, not ONLY to make, but the Mega-Staple Machine, that drove these massive Earth Staples! After 22 years in Construction, I can say that Humans never had this kind of massive Work imagined...with the exception of Science Fiction Books on...Terraforming.

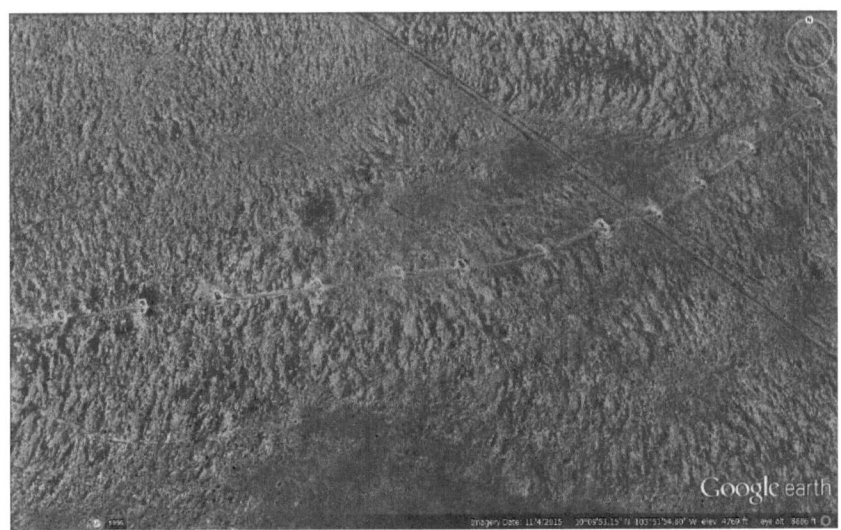

Fig 92

STAPLES AT REGULAR INTERVALS IMPLY MACHINERY

Obviously, this was intended to keep everything in Place, even through Earthquakes and Flood! The Size and Scope of this undertaking, is truly Megalithic, and utterly Mind-Boggling! The People, who were responsible, could Truly mold the Earth, like the 'gods'.

As Steven Hawking once said, (paraphrased) 'Any Technology...at this level...would seem like Magic'.

This does not mean that everything is beyond our 'Ant-Brain'. There are Structured Areas, and, what appear to be, Burial Mounds, Pyramidal and Rectangular, obviously eroded places, that almost always indicate this was much more Populated.

Fig 93

OBVIOUS SIGNS OF STRUCTURES

Fortunately, Texas Laws do NOT allow Claims to be filed, so I don't expect that release of this information to have the Plateau run-amok with Treasure Seekers. Hopefully this Book will help stimulate some Credentialed Archaeology.

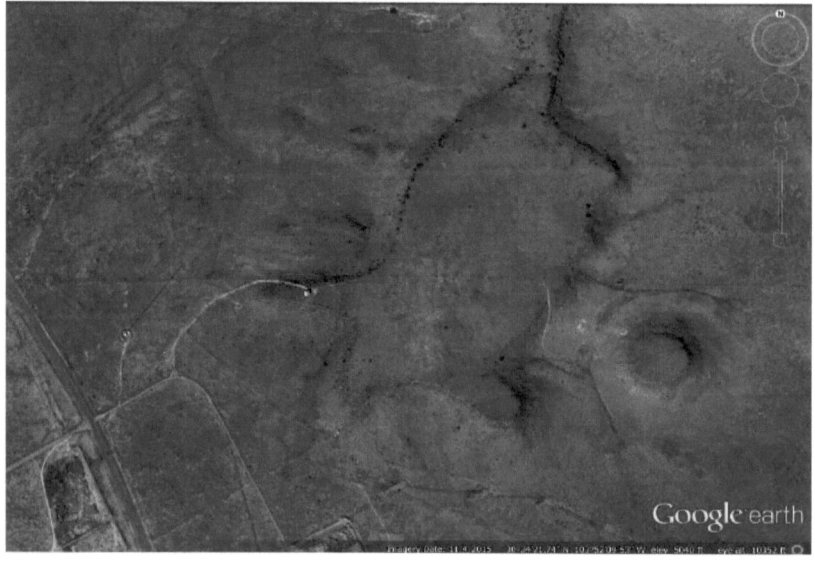

Fig 94

UNNATURAL FORMATIONS ABOUND AROUND PLATEAU

Fig 97

HEXAGONAL CRATER WITH RECTANGULAR TOP

Although the term; Pareidolia, is itself, a 'Pseudoscience', which 'real' Scientists would like to explain much of 'Paranormal' sightings with, I don't claim anything but the Structure itself, which is, at the very LEAST, 'Anomalous' in, and OF it's own 'impossible' Formation!

15. THE ANUNNAKI CONNECTION

This picture, is also one of the inner connectors, obviously has someone living in the area, in question. I believe this is a multiple 'Face' artwork, meant to depict three Faces; there is a Simian face, connected to a double Face, which depicts both, Anunnaki, or 'elongated' Head, and a Human Face. The Anunnaki and Human, are 'under one Hat', and when slightly zoomed- out, all three faces connect from a line marking the Human Eye, to the Simian image. Also of note is, what appears to be a box with a handle, depicted. Whenever the Sumerians depicted the Anunnaki, working with DNA, they always had a little Suitcase! Again, just too many 'coincidences' for Statistical Coincidence.

Fig 98

SIMIAN CONNECTIND TO MIXED ANUNNA-HUMAN FACES

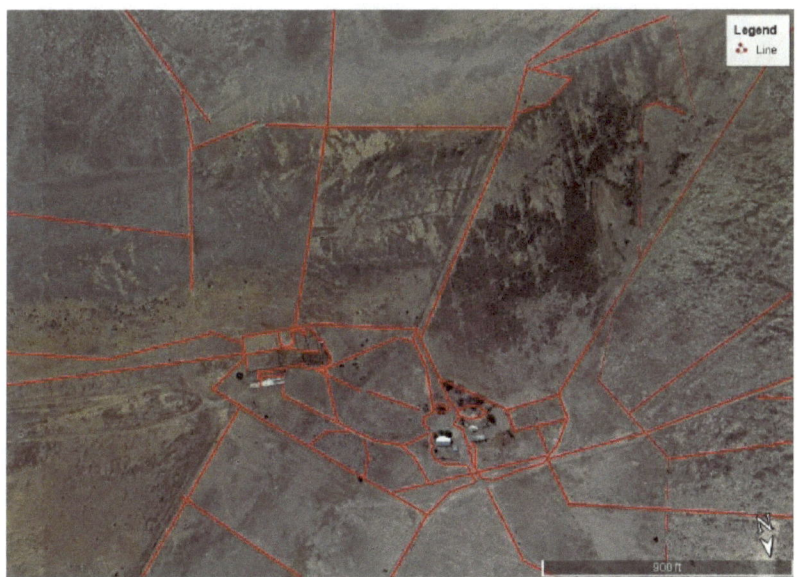

Fig 99

ZOOMED-OUT SHOWS ANUNNA-HUMAN UNDER SAME 'HAT'

I found many of these, Multiple-faced depictions. All of them indicate that Elongated Headed People, are connected to both, Human and Simian species, somehow! Brian Foerester, has been studying the Elongated Headed Species of Bipedal Hominid, in Paracas, Peru, for quite some time. He also has been doing DNA Testing, on some of these non-human skulls. His work may also validate the Sumerian history of the Anunnaki, which indicates that Human Beings have a Matriarchal connection, to this Species!

Fig 100

MITCHELL FLATS AREA ORIENTED TO STONEHENGE, UK

If you will note, the orientation is the same as the Matriarch figure, shown earlier, a little bit of Research indicates that THOTH was Honored, in Statue form and Papyrus, a Baboon! I say 'Honored', because there was no stigma associated with this Symbolism Egypt.

Also, of note is; this is in an area of much activity, in the Marfa Lights, called, Mitch Flats. Apparently Marfa has regular Tours, to this area, for Night Viewing.

Most of these multi-faced Images, are found by their Eyes, which are a hallmark of the present-day, Grey Alien, and Abduction phenomena. Once you see the Eye, it's easy to connect the Image; they all seem to have a certain square-ness to the Jawbone, and of course, the smallness of Nose.

The Sphinx, previously shown, although there are signs of extreme Erosion, has, what I believe, is another representation of an Elongated Head! I was unaware that the same representation, comes from Egypt, in the form of the oldest Image of RA, landing his Craft on the back of the Sphinx at Giza!

Indeed, there are even signs of an Archway, on the back of the Sphinx's Neck! Evidence of a VIP entrance to underground Giza? Unsurprising, that Egyptologists don't publicize that fact.

It seems that there is a whole Tunnel 'System' that Credentialed Egyptologists, don't want the General Public to know about. Again, it parallels, exactly, what is written in the Emerald of Tablets of Thoth, regarding the Giza Construction, and, again, is completely dis-regard by Egyptologists. Real 'Truths' don't need belief, or acceptance from Educational 'Peers', they simply exist...like the Tunnel System they don't talk about.

Remember, this is all connected to the Path from the Apex of the Great Pyramid at Giza, passing exactly over the Altar at Stonehenge!

Fig 101

OLDEST PAPRUS DEPICTING CRAFT LANDING ON BACK OF SPHINX

Fig 102

NIMROD LOOKS AT ANUNNA/HUMAN/SIMIAN CONNECTORS

The last picture I call, Nimrod, because of the type of beard, he appears to sport, distinctive in Sumerian Artifacts. The Face is a Giant Geoglyphs, that measures Five miles, in length, by just over Two miles, in width. The Line from the Eye leads directly, to the Multi-Face, at the beginning of this Chapter.

All of this evidence has led me to where I began: the Anunnaki again! AND, if you are going to do any Serious Study, of the Species of gods from the Stars, your Research will bring you Mr Gerald Clark. As mine did.

Like myself, Mr Clark is not a PHD in Archaeology or Anthropology. Unlike me, Gerald Clark has Degrees: a BS in Computer Engineering, and MSEE in Electronic Circuits and Systems (Masters in Electronics), has two Patents, was a Helicopter Pilot in the US Army (1982-89), and happens to be the Author of two Books. The Anunnaki of Nibiru, which clearly establishes links to documented History, at a Worldwide level, and The 7th Planet, Mercury Rising, that follows Enki's Son, Thoth, and their effort to Free Mankind, by Knowledge, and the Emerald Tablets, as a Key, to that Freedom.

Mr Clark is, in this Author's Opinion, the best source of Information on all things, Anunnaki. He is also one of the NICEST people, I've had the pleasure of having intelligent conversation with.

I had written Mr Clark, about the Lines and Glyphs I had found, to which, he not only validated their Presence, but brought to my attention: that the Great Solar Rays, were likely indications of Atlantis! I looked at the main Solar patterns, and simply counted them...Seven in total, Six, that surround a Central, or Seventh Symbol, the Number Seven, being indicative of Atlantis, Enki, Thoth and the Sumerian Number of EARTH!

16. HYPOTHETICAL UNIVERSAL COURT CASE

Firstly, you have to consider that there IS a Race of People who are named 'Anunna ', that come from a Planet THEY call 'Nibiru'.

Secondly, you must accept the Sumerian Historical Records as; accurate portrayals of this Race of Peoples, and their History, on this Planet. Remember that the Anunna called this Planet; KI, and the ensuing Progeny were thus known as; Anunnaki. Those were Anunna BORN on the Planet: Ki.

Thirdly (and most importantly), you must consider their Technology is AT LEAST, a Million years, or so, AHEAD of anything in Human 'Sciences'. Why? Simply because THEY were responsible NOT only for; the Creation of Earth, Humanity, BUT also the 'seeding' of all other Life on this Planet. Quite possibly, the Anunna have had the Time, Motive and Opportunity, to be the Ancestors of any OTHER Race of People, in the Galactic Area.

According to the Sumerian Chronicles, the Anunnaki have had regular, and sometimes catastrophic, Planetary orbits in Sol's Solar System, long ago, one of their Moons collided with a Planet, Orbiting between Mars and Jupiter. The remnants are known to us as the Asteroid Belt, and Scientists know that there isn't enough material there, to be a sizeable Planet. The Sumerian Record goes on; there was a large piece the Anunna called Tiamat, that they placed in a closer Orbit, which they named: Ki, and proceeded to Terraform, and Mine for the Gold. Apparently they had problems with their Planet, and Radioactivity. They had tried several methods of Shielding, and found that Gold had Properties that they desperately needed for Atmospheric protection.

But the Work of Mining was too much to bear for Thousands of years, and the workers rebelled when they found they could never return to their Planet, they had truly become part of the Earth: Annunna-ki. They became conscious of their Slavery, apparently now 'illegal' on Nibiru, was found 'acceptable', based on the larger need on the Home Planet.

The rebellion was violently put-down, and the 'god' who had conspired was killed. It was decided that they would USE their own DNA, infused with part of the defeated 'god's' Consciousness (so they could understand Orders), and, after several tries, came-up with 'modern' Human, to be a kind of Worker-Class for the 'gods'.

At some point, the Anunnaki had Mined enough Gold, and one of the main god: Enlil, had petitioned HIS Father, the Leader on Nibiru, Anu, for an END to these noisy Monkeys. Anu then Orders his '2cnd' Son, Enki, who was in 'charge' of Earth's Waters, to wipe the Slate.

Enki and his Son, Thoth, who were the key Scientists in the Creation of Humans, Conspire to save a portion of Humanity, but bring the Flood that was Ordered by Anu.

So starts a series of Wars, between these Brothers, stuck on Earth, one of which despises it. They continue their DNA/Consciousness Science, one seeking to make Humans their Intellectual equals, one seeking their end.

Lately there have been a few Authors who seem to think there is some kind of Gala Authority over ALL the Space-going Races in the Area, some have even named them; Solars, Wardens, or, Watchers.

Let's hypothetically consider a Galactic Court of Law:

CASE: Humanity V Anunna-

FOR: Violations of Slavery, Corruption, Coercion, Murder, Genocide, Environmental Destruction

Judge(s): We've read the Charges against Anunna, and have allowed them to proceed, unfortunately, since all other Races in the near Galactic Community have traced Lineage directly to Anunna Genetic Manipulation, none can be regarded as 'Peers', thus, no Jury will be selected, and any Judgment leading to Punishment, against Anunna will be on a Voluntary basis. HOW do you Plead?

Enlil: Your Honor, I will speak for my Father, Anu, and my Race. We plead Not Guilty to Property Laws, we seek full dismissal of Charges.

Judge: Since the Galactic Community regards Knowledge and Conscious Awareness as Universal Commodities, Individuals or Races, that have reached a Stage to Contribute to the Universal Quantum, or, to affect Creation, are considered as PART of this Galactic Community, with ALL Community 'Rights', and Laws Due. At that point, they can NO longer be used as Slaves. Can you explain your actions?

Enlil: When we created Humans, WE Elevated their Consciousness, using the Quantum from our OWN People, at the time of their supposed Slavery, we had Programmed their DNA to SEEK & VALUE the Gold that we needed, and to take JOY in their Service to Anunnaki also placed several 'Kill-Switches', to ensure they would never reach full Potential of Quantum Awareness. When they were not as successful, I ordered a World-Wide cleansing but my Brother Enki betrayed our Father, and the results are now Suing. At the time, they were not considered capable of adding to the Universal Quantum, so, in essence; we did NOT Enslave, Murder OR commit Genocide against a Race of Consciously aware Beings.

Judge: WE must agree with the Honorable Enlil, that the Human species was incapable of Consciously affecting Creation at the Time, and Formally DISMISS the Charges of Slavery, Murder and Genocide. Still OUTSTANDING are: Corruption, Coercion and Environmental Destruction. Will you answer those Charges?

Enlil: After the attempted Cleansing, we too, saw their Potential of adding to the Universal quantum, and our own research led to our being able to incorporate our own Quantum into their daily lives, thus allowing Anunnaki to continue their Life Force, in a Human Body... At the Present time we ARE attempting to steer their Consciousness by forcing them to live in Primitive states, as all must live in Physical States. We have encircled their Planet with Radioactive Belts, to further teach them, through isolation. As to the Environmental Destruction Charge; we created this World, and

Environment, from a disaster we did NOT create, we have NEVER Vacated this Planet, my own Brother still lives in the Flower of Life that shines as the very Heart of this World, our 'Masters' have found a kind of Life, as the very Attributes, that Humans use in their Daily Lives. Even now, we are attempting to 'correct' Kill-Switch in Human DNA, and much of Ki's environmental problems are simply due to the Fact that Humans are still un-ready to join the larger Universal Community, at this time, and Cyclic Climate Change is wreaking Havoc, Humans are just now learning to control the Weather, and have added to the Cyclic problems.

Judge: Although the Anunnaki, seem to, I find that they have NOT abandoned their Property, but may NOT 'evict' any living material from Property, because of Species Enmeshment, this World is to be prepared for full involvement in the Galactic Community and ALL Charges are formally Dismissed...Court Adjourned.

Enlil: Thank You Your Honor.

If only a Hypothetical look at the Proprietary nature, the Anunnaki would still have MORE of a 'Claim' to the Earth, than Humanity, even without a 'Court' to Rule, if you accept the People that were Chronicled by the Sumerians, as a FACT, then you also can see an inkling of THEIR view, of US!

I can't say that I know how another Species thinks, but evidence included in their Chronicles, tells of Incest, Murder, Political Intrigue, and even Wars fought between the Anunna, and AnunnaKi. They would Easily Pre-Date our Species, so our tendency to Violence might just be Genetic.

There are many Authors who are better qualified, than me, to learn about the fact History (if not purposely hidden) of the People the Sumerian knew as; Anunnaki. Other Peoples had other Names for the same Beings: to the Hopi, Enki was known as; TAWA, Isis, the Egyptian god-Mother, was called; SPIDER WOMAN, many other of the 1rst Peoples knew them as; OLD MAN, and OLD WOMAN. If you picture a Small, Grey, Elongated headed Person, you can easily imagine comparing him to an elderly Person. Remember, AGE denoted Wisdom, and demanded Respect, especially to a short-lived Species.

My Friend, and incredibly Talented Author, Gerald Clark, The Anunnaki Chronicles, and The 7th Planet-Mercury Rising, would be my Personal suggestion for up-to-date Anunnaki information. You should also remember, none of their History would be known, without Work of Zechariah Sitchen, who handled his Critics with Fact, and Translations of Sumerian Cuneiform, which are now available online, for ALL Seekers.

17. MARFA PICTORIAL

The following Photographs are from my Personal Trip to Marfa, April 2016. Taken from the First Public Fly-Over, at Altitudes that these Lines and Geoglyphs can be properly viewed. Although Others have, no doubt, seen them, Others haven't seen them for what they are...

From Top Left:

RV by Share My RV, Inc

Good Place To Stay In Alpine, TX

Frontier Mason Temple, FACES Marfa Plateau

(Unclear WHY Needed for a Small Population)

Marfa, TX

Tourist Info at Former USO

 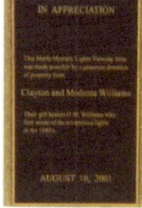  

Top Line: Downtown Marfa 2016, Marfa Lights Viewing Center & Plaque

Middle Line: Binoculars are Public, Commemorative & Informative Plaques in Viewing Area

Bottom Row:

Travland Helicopter Intnl, Inc.

PO Box 2019

Alpine, TX 79830

Ph 432-837-1848

Great Pilot; Lewis Travland

Great Helicopter

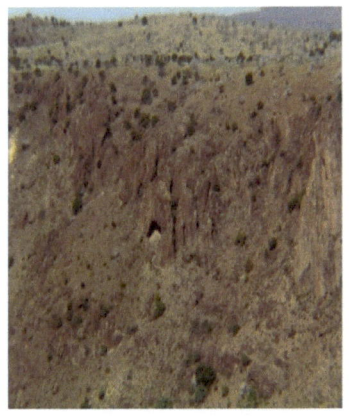

From Top: Caves Abound on the Way, 1st Lines easily Visible at Marfa Plateau, HOW Raised Lines Look from Ground and Aerial, Dark Lines are Embedded in Soil & Structured, Lines Cross Hill & Valley Perfectly Straight.

From Top: 3 Solar Glyphs, apparent Metal Dam above Marfa, The Next Photos are of the Matriarch's Eye and Clearly Show Raised Lines, Some Color Enhanced, Carved Lid on Eye?

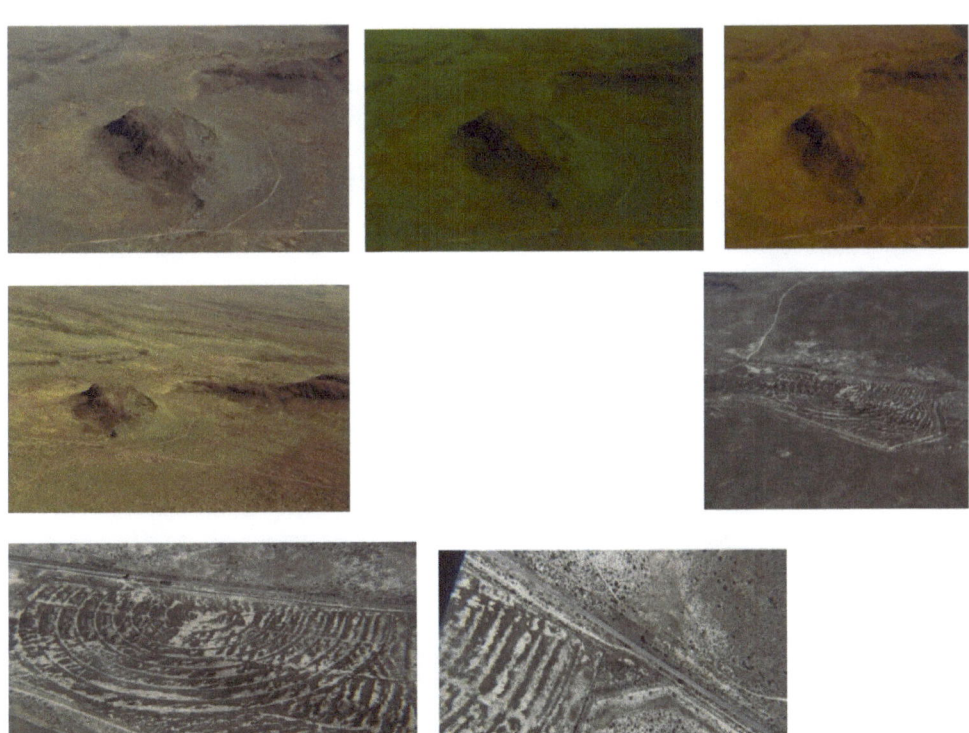

I Hope the Color Helps See the Archaeological Evidence Fixed-Wing Aircraft suggested for more Altitude.

Mitchell Flats, Famous for Light Activity, are made of Raised & Etched Lines, does NOT reflect Circular Farming.

Metal al Staples Infer Terra- forming Dark Lines are NOT Fencing, Quartz Crystal in ALL Soil over Plateau. Let the Next Pages Speak for Themselves.

PLEASE HELP Save This Precious Area of History, FOR THE FUTURE!

18. CONCLUSION, SPECULATIONS AND INFERENCES. WHY?

The Eternal Question of Consciousness.

Remember your Child asking; "Why is the Sky Blue?", or maybe it was you, asking Your Father or Mother? If you do, then you know the question is literally, unending.

It's what has driven me to pursue the Evidence I've found, and the resulting connections, I've followed. Each piece, leading to the next, like a Worldwide Jigsaw Puzzle, the Pieces only fit, if placed in the correct order. Otherwise it looks just like the jumbled mess, that modern 'Academics', have made of accepted recorded 'History'. Almost like the Ant on the Mona Lisa metaphor, they're just too close, for accurate perception.

Let me explain; true Archaeological discoveries have Egyptian Artifacts, being found all over the World! Other discoveries show Technology existed at a high level, far in the Past, so far that, if Human Evolution were true, then Human Beings were not the Makers or Users, of said Technology.

Notable Archaeologists, like Michael Cremo, has Lectured and Wrote, extensively, about Million Years Old hand-worked, or Machine manufactured Artifacts. Some of which, were found in Coal, or surrounded by Fossil material! If you don't understand the significance; Coal takes millions of years to form, as do the circumstance of bone turning to Stone, thus, fixing the approximate Dating. AND, Scientifically Dis-Proving the Theory of Evolution, along the way: if only ONE of the thousands of such, 'Out of Place Artifacts' (OOPART), are real!

If you need more, of the things that 'Evolutionary Biologists', are lacking; otherwise known as Evidence, we can now Peruse the Scientific Library of the former Soviet Union.

You see, back when they were a Socialist Government, they had plenty of Money, to use for Pure, or, Theoretical Science, and was then tested in Physical Experiments.

One of which, involved Drilling a great, deep hole. Originally, they intended to reach a depth of 33Km (thirty-three Kilometers), but stopped, after reaching approx. 12-13 Km, due to loss of funding. But, along the way, they had found several items, of Scientific interest: at that Depth, Rock becomes Plasticine, and the World's Oldest Fossils, or Plankton, dating at least 6 Million years, and basically unchanged!

It's a fact, that Darwin, himself, wrote that (paraphrased), 'This Theory should be Dismissed...', if proofs weren't met! Those Proofs, included discovery of Transitional Fossils or Fossil Evidence, of Species-becoming-Species, and there should be lots and lots, of them, but they're not. Such Notable Evolutionary Scientists, as the Hon. Richard Dawkins' pontificates, 'Whales...became Elephants, who...didn't like it and went BACK to the Oceans, to...become Whales, again' (paraphrased).

Every few years, or so, one of these learned people, will find some fossilized Bones, and claim they are Mankind's distant Relatives. It happened last, when they, very Publicly, paraded the Lucy fossils, found in 2004, as The Missing Link.

Of course, everyone remembers the Movie, filmed in 2009, but, what you don't remember is, the Bones were proven to be Monkey Bones in 2007!

Apparently, you are real Science, if you are based on lack of evidence, lies and outright Fraud.

Of course, when whistleblowers in Governmental Agencies, give a Document stating; (paraphrased)'The question of Human Evolution...on this Planet has been asked and answered', to a Hero, like Linda Moullton Howe, and moreover that, ' This Government...is no longer interested', it is expected to become Public knowledge!

According to the Fictional Detective, Sherlock Holmes, "Let the Evidence speak for itself", and, if the evidence leads elsewhere, you must, "Follow the Evidence"

That Evidence has lead me to the only Logical Conclusion:

The Anunnaki are real.

Their History is Chronicled, in Cuneiform Tablets, Scrolls, Artwork, and Cylinder Seals. Their History is depicted in the ENUMA ELISH, and ATRA HASIS, and other Stories that are the Oldest known Written Human Histories!

The Sumerians had the first known form of Writing, in which they kept a detailed Report of all the quantities of material and foods, the Day to Day workings of their Government,Leaders and Gods from the Sky!

Sumerian History also details those gods created everything, for their own purpose! I won't be Teaching the History of the Anunnaki, except as paraphrased, because I believe those are better told, by someone who has specialized in that Field; Namely, Gerald Clark, who I have personally found to be the best, most accurate and interesting source. Clark not only follows them around the World, but brings the Key to true freedom to light in his Book; The 7th Planet-Mercury Rising, and the Artifacts known as; The Emerald Tablets of Thoth.

IF the Emerald Tablets are REAL, then Anunnaki are very familiar with a kind of Consciousness Technology!

If the Anunnaki are real, it explains, well,...EVERYTHING!

I submit that the Elongated Headed Species, whose Bones are found throughout the World.

WHO ARE THE ANUNNAKI? FOLLOW THE EVIDENCE:

1. Anunnaki are linked to Stonehenge, United Kingdom, where, very recently, Elongated Heads, with skeletons, were un-earthed in close proximity.

2. Stonehenge, similar Shrines and Structures are linked, shrine-to-shrine, always including at least one Pyramid, by perfect alignment of Summer Solstice, and Winter Equinox Paths.

3. Stonehenge, and similar Megaliths, are linked to the hidden Phenomena of Giants.

4. DNA evidence are linked to the Paths that Peoples followed, across the Earth, and can be matched to varying Solstice Paths.

5. Megaliths are linked, by Polygonal Construction similarities, and therefore link Anunnaki, Giants, Stonehenge (and similar structures), by following Summer Solstice, Winter Equinox Paths.

6. SUN, or Solar Religions, are linked, to all of the Phenomena listed above, by virtue of the sameness, of the Characters, edifices and Text.

7. The Discovery of Lines and Geoglyphs in Marfa, Texas, are linked, by following perfectly straight Path, which links Stonehenge, and the Great Pyramid at Giza! (don't forget Aden, in the Sinai Peninsula)

8. The Story of Atlantis is linked, to ALL of the above!

Even, the Great Fictional Detective; Sherlock Holmes, would conclude that, not only everything evidentiary connected, but, that everything is meaningful to the People, who builtit all!

Anunnaki, connected all their Shrines and Pyramids, in perfectly straight Lines.

I can only think of two possible answers: 1- Communication, or 2- Transportation.

Solstice Paths cannot be seen. Linking the Shrines for Communication doesn't make sense; Radio Waves expand over distance, and are a very inexpensive form of communication. So, the need for such precise alignment, to communicate, would mean a Laser, or directed communication via Light. I discount this idea because of the Curvature of the Earth, they would need reflective material, at regular intervals, not in evidence.

That leaves only Transportation. According to Sumerian accounts, Anunnaki came for the Gold, and, considering the weight, must have needed places of Refinement, to get it Off-this Planet. Thus, we come from Shrine, to Pyramids, for storage and refinement.

The idea that Pyramids are places of Gold Refinement, is not new. This Theory has even gotten some, grudging, Academic support, and new discoveries, under the Pyramid of the Sun, Teotihuacan, Mexico, have located a "lake of Mercury", an Element commonly associated with Gold Refinement! I would personally call this evidence.

The Story of Moses, feeding his followers Manna, after fleeing from Egypt and the Armies of the Pharaoh, infers the conversion of a Golden Calf, into a form of food, then forcing all to eat. This has been called Monatomic Gold, which has properties Anti-Gravity, among other anomalous attributes, like the changes that take place when ingested!

Wouldn't a superior Intelligence realize that; converting Gold, to Monatomic Gold makes it much easier, cost-efficient, and ready for dispersal in Atmosphere...the chronicled reason the Anunnaki were here!

It must have been some sight; watching Lines of Gold, leaving the Great pyramid through the Small Tunnels that point to the Constellation Orion, coming from both the and Queens Chambers. All the while, Lit-Up like a Light bulb!

You see, the Discovery of Nazca Style Lines in Marfa, TX, infer that the Plain of Giza was Electrified, using a form of Hydro-Electricity! If True, then the Theory of some kind of resonant, or Pyramid Energy must fall, because Pyramids are Energy Users, NOT producers of Energy.

The Limestone underground at Giza, is a huge Aquifer, although much depleted, Water still flows. Tunneling is extensive and, undeniably complex, and shows the passage of water through them. The Plateau of Giza was levelled-off, perfectly, then huge Polygon-ally cut blocks were fitted over the whole Plateau, and finally, the Edifices were built. Once you accept that the Aquifer was Electrified, everything else falls into place.

I've shown evidence that the Chinese succeeded in electrifying the Polygonal Lines I found (by Solstice Path), in the Gobi Desert, by apparently running Water through an existing Canal system! But, not shown, is the effect you see when zooming-out: there is streaking visibly brighter, surrounding the Canals they infuse with Water! Somehow, they are electrifying the Aquifer, and it acts like a light bulb, lighting the Soil, above, which infers; Quartz. The Co-ordinates are previously given, for review.

Now, I had evidence of a Quartzite Cap, over much of the Nazca, Peru Glyphs, pardon the Pun, but, the connection was obvious, to me; both the Sites of Geoglyphs, Nazca and Marfa, TX, were meant to be visible from Altitude, and Lit-Up as HUGE Light Shows, from under the Ground!

It infers that the Term: Dark Ages, could have been a real Worldwide phenomena! When the lights really went out, it's not that big of a stretch, that Thoth, tried to help the Hum their Children, retain Knowledge, by Religious Ritual, to guard against the coming Dark.

Which makes the Marfa Plateau, quite a bit more interesting than Nazca, is that some of the massive Geoglyphs I have found, seem made of different Color material, in order to stand out from it's surroundings.

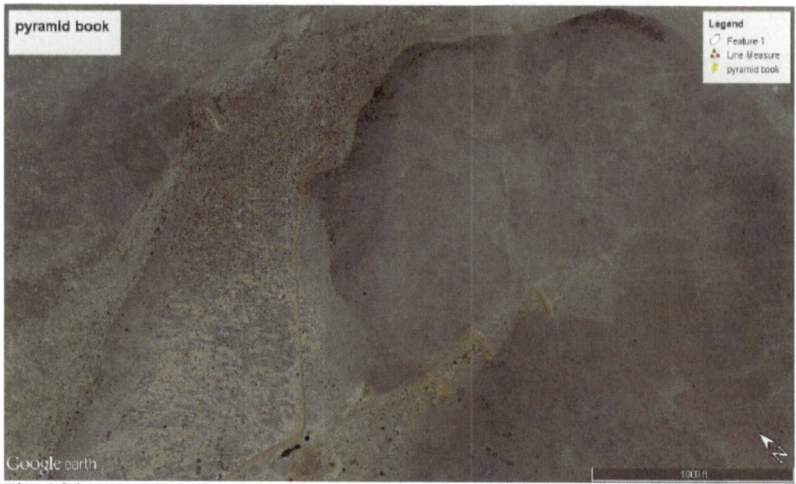

Fig 103

PRIEST

This is a feature I call, The Priest, note the decoration at the neckline, this appears to be metal! The Image is surrounded, or emphasized, by a Line system, and is also oriented along the Solstice Line passing through Marfa. Color enhancement clearly shows, what I think, is how it looks when Lit-Up from under the Glyph.

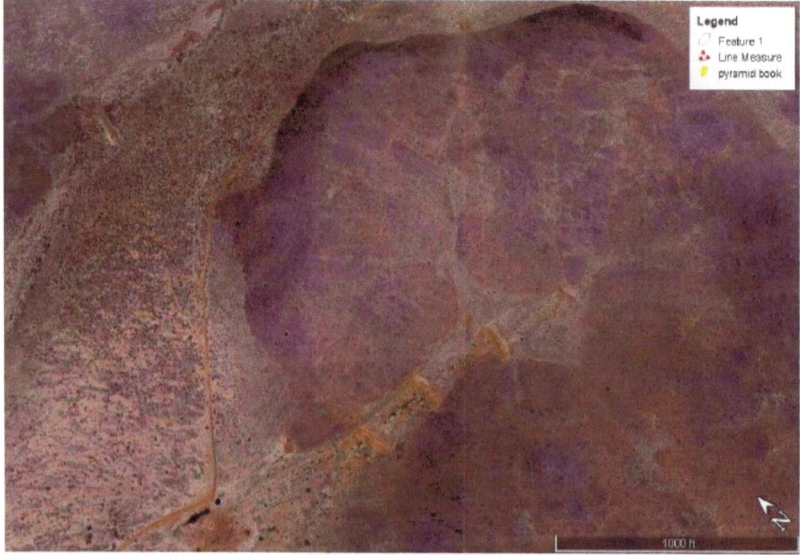

Fig 104

CONTRAST AND COLOR ENHANCED PRIEST

To quote Mr Hendrick, when he viewed the Marfa Lights in their active stage, they looked like "a Three-Ring Circus". And you also should realize that, on a dark Planet, areas of Light and Activity would certainly be helpful, as you flew above, for Navigational needs. With an Artistic Flair, no less!

Now, you can imagine what Giza truly looked like, and Nazca! When you realize the Earth truly is differing structures of Quartzite, and you add a Species, who works with Stone, differing Color, for Artistic Purpose, mixed with functional design, done with an ease that boggles the Mind!

The Ant Metaphor strikes again! But just because your insect Mind can't understand the Science of Mona Lisa's Painter, doesn't affect the FACT of the Painting's existence!

Polygonal Construction, can be explained by a very recent Translation of a small Sumerian Tablet; it carefully details the movement of Jupiter, through the Night Sky. The Translations show a different form of Mathematics, than modern science: we have developed a method of fixing the location of a point by Triangulation, called; Trigonometry, Mathematics based on Triangles. Sumerian Mathematics is based on Trapezoidal shapes and (at least) four Points in space, and their methods of Construction would follow this Pattern, not associated with Human Construction, through history.

Following this evidence, connects the Sumerians, thus the Anunnaki, to that unique Construction Method. You might note; that Anunnaki Math would be a much more precise method of determining locations in Outer Space!

Evidence further shows that some kind of Electricity causes the Plasma extrusions, known as Marfa Lights. The Similarities of Construction of Nazca, Peru, and even the Giza Plateau, itself, cannot be denied when identical methods are visibly recognizable.

I speculate, that Polygonal Construction Methods, might have helped regulate and direct the Plasmic Electricity, they were Generating.

Many people look at the colossal size, of the Blocks, involved in the Mayan Walls, and they notice the Puckering, or Shaping, that looks as if the Rock were, somehow Fluid, at the Joints. But, to a People coming from a Planet of higher Gravity, and pressure of Atmosphere would have known, probably a common-knowledge to them, that Rock, even the hardest, becomes Plasticine under pressure. Remember the Soviet Union's great deep hole, I told you about earlier?

By the way, the Deep Hole Experiment of the Soviets had a couple of other interesting results: they found that Hydrogen and Oxygen Molecules were being forced together, to form Water (H_2O), at deep pressures. This infers that any World, with hydrogen and Oxygen in an Atmosphere, will form Water in Aquifers under pressure of sufficient Gravity!

One more thing; Soviet Russia also found that Petroleum, is not Fossil-Fuel! It seems to be Coal, that Liquefies under pressure, forming in the same sort of 'aquifer', as Water does.

Now you know how both Oil and Water are formed, by Natural Process, and are Abundant, which means these things are Naturally Replenished, and should NOT be thought-of as if we are running-out, of them.

Plasma Physics, is a fairly new development, in the field of Physics, and could be the real explanation for results that happen in the famous 'Double-Blind Slit' Experiment. That's the experiment that shows a Single photon, of Light, aimed at a wall, cut with side-by-side slits, will pass through both slits, suggesting Waves of Light (Plasma). Unless, of course, you Consciously View the experiment, in which case, the Light will pass through one, or the other but not as a Wave, but exactly like a random pattern of hit-or-miss, particles (think of a Pitching Machine, and particles, as the Balls). Which suggests that Consciousness affects reality, or, as Albert Einstein once said, "We Create our reality".

The Proof of Plasma Physics is the fairly new discovery of Galaxies that exhibit two different Red-Shifts! It's now roiling the Academic Halls known as, the Red-Shift Controversy.

Upon close inspection, the varying Red-Shifted section is; Quasars, ejected by the Galaxy, are attached by visible tethers that are transferring, or feeding the Quasar, some kind of Plasma! Like a Galaxy giving Birth, to a new Galaxy, formed when the Quasar reaches another stage of development!

This infers that, once the original Big Bang happened, the pattern was set for many small bangs, which explains the seeming motion of everything moving apart, without the Magic Dark Matter, that can't be located. Also being inferred, the fact that Red and Blue Shifts might not be an accurate method of determining distance.

The Universe then works as a Function of Light in a Quantum State, or, all kinds of different Plasma is moving through the Universe, at different vibrational Levels, which creates Quantum Entanglement, we know as physical reality. And, Quantum Entanglement infers some kind of Conscious Universe!

The Atra Hasis, Enuma Elish and The Emerald Tablets of Thoth, give a recorded History of Anunnaki use of Consciousness Technology. Again, higher Truths NEVER require belief to be valid; Truth, justifies itself.

Best Example of this truth is, the Great Pyramid itself: obviously, no one needs to believe this is High Tech, for it to Physically express it!

I would suggest that any Research of Sumerian History begin with someone who has specialized, and carefully re-constructs the Worldwide Phenomena of Anunnaki, and their true implications, like the Author, Gerald Clark.

UFO and Abduction Phenomena, can all be attributed to the Anunnaki. Although there is anecdotal evidence of other Species that have visited this Planet, the only real evidence of Non-Human Species, other than Giants, IS of the Elongated-Headed Species, that are visibly the ONE species recorded to be involved in the Abduction Phenomena, the Grey Aliens.

When I have been able to ask Experiencers, as they are now known, if they got them to explain why they were Taken, Experiencers invariably say, 'They said they were...correcting DNA!'; something the Sumerians detailed, when they created Human Beings! Medical evidence that some Experiencers have been cured of illness, even Cancers, is abundant.

Then, there is the evidence of Splicing, in Human DNA, NO OTHER Animal has Junk And acts like some kind of Programming, in which, Cancer is found, like a Kill-Switch, that has been linked to Radiation being a Trigger! Truthful Oncologists, will verify this.

Solstice Paths, are not Visible, so I am Speculating that they were a function of the Navigation needs when going from Altar-to-Altar, picking-up Gold for Transport, which were both Refined and Stored-at Pyramids!

Anunnaki then Taught the importance of following the Sun, setting-up Religions, so the Humans they created would Revere the things THEY did!

In the Emerald Tablets, Thoth even says he did this so we wouldn't forget, to 'make a Ritual of Knowledge', because he knew, the Darkness was coming! It probably explains why the great Religions, all have the same Characters, Roots, and Mystic Knowledge. Exceptions are the Religions of the Orient, which Teach the hidden Knowledge, more openly; namely Taoism and Buddhism, express the relationship between the Creator and Created, without secreting them in Parable, as in the Torah and the hidden Numerical values, embedded in Ancient Hebrew.

And, if the Ten Commandments come from Anunnaki, then 'Honor thy Father and thy Mother', is probably mistranslated from, 'Honor THE Father (Enki) and THE Mother (Isis)'. Egyptian/Sumerian creation of Human, Story.

The Marfa Megalith represents ancient Technology, that still works! Also represented there are Geoglyphs that picture different Religious Symbols: Egyptian, Cross, Star of David, Moslem Crescent, and Taoist Yin/Yang circle (others too), images of Animals, with Mayan influence, pictures of people, including Olmec, African,

Caucasian and Oriental too! ALL intended for viewing above the Earth!

Atlantis is represented in Marfa, so this is likely a representation of Atlantean Technology too!

Upon reviewing Mr Hendrick's experience, his equipment was set-up to detect Electromagnetic sources, which he did not. What he saw, were visible Plasma extrusions, both Timed and Regulated, as a controlled event. I speculate that the 'whistling' effect equipment did detect, was Plasma Electricity, moving through the Connective material. Soil samples taken at the Lines should prove or disprove this. As would shooting a Charge in areas of reported activity.

I remember that a Lawyer once used a Phrase that goes something like, ' There is ZERO Percent possibility, that there is NO Cause and Effect'... to the existence of the Lines, Visible Electric Activity, and Art, made from the Earth being Manipulated, to make the POINT, of being viewable from Altitude!

If you read the history of Anunnaki, as recorded by the Sumerians, they come from a Planet, much larger than Earth, and had problems that threatened their Species with radiation. If you Imagine a Species living in such an environment, they would probably be smaller in Stature, than Humans, their Bones would be very Dense, from the pressure of their Atmosphere, and the added Gravity of a larger Mass. The Radiation problem would drive them to be an underground adapted Peoples, and would likely have Vision adapted. And, that Big Planet that NASA just found, reads just like Anunnaki's home; Nibiru.

If all this sounds familiar to you, you should realize; I never intended to find the Answers to life's Questions...in Marfa, Texas. But then WHY not? Texas is a BIG State.

According to the Emerald Tablets, the Anunnaki, have the Consciousness Technology that allows them to, ' Live in Man', or to Incarnate, in Humans. If you realize that the vast majority of People, believe in re-incarnation, in one form or another, and you add the hidden Judaic version of re-incarnation, the numbers go way higher.

The Anunnaki Masters, who apparently still live in the Halls of Amenti, in the Middle of the Earth, live as Attributes, in Man, when they want. To realize that our Souls have a Path that stretches to Galactic measure, is Mind-Boggling!

ONE more discovery; according to Thoth's Emerald Tablets, ENKI, his Father, still lives in the Heart of the Earth's Flower of Life. And, as Yeshua (Jesus)/Thoth said, 'No one goes to the Father, except BY me', can be read: None reach the Creator without Knowledge!

Letting Inspiration lead me, I placed a Layer, over the Stonehenge Blueprint, of a Flower of Life Symbol, and found: they MATCH! Then I did the same, at the Great Pyramid, and with the same results, which leads me to ONE last Speculation: The Anunnaki modeled the Planet Earth, on the Flower Of Life. THOSE are the Paths they follow, as Orion himself, whose Shoulders Point to the Solstice, and Legs to the Equinox.

ALL Geometric forms, can be found in the Flower Of Life, As Above...So Below...and more to the point; if Stonehenge is modeled after the Flower of Life, the Sacred Geometry expressed, is the reason for the amazing Mathematics, in Stone.

What Marfa, Texas has shown me is; that there was a REAL Time in History that included all the different Peoples of Earth, working together, for a Common Goal. There was a One World Government, whose Rule was Totalitarian, but who revered Spirit and Knowledge, and did NOT separate Peoples BY the Religion they called-on.

Oh, and if Marfa was built to be a Light Show, then I have to Conclude that the Car light's now on some of the Lines, only are a pale semblance of the Plasma, that once zipped up-and-down them. So GO and view the Lights, whether made by Cars or Atlantean Technology! We can Imagine them as the Plasma, once, and possibly still Visible!

And the real Evidence of these Lines being present: at NO time till the Present, in recorded History, has Mankind ever made Roads, meaningful at Altitude! (maybe Flying Ants?)

I've heard Politicians want to build a Great Wall to keep other People on a particular side of Border. I've found that these Lines, and Glyphs don't know a border, so a Wall would only serve to continue to hide, what surely belongs to all.

I feel I must note; at the Publishing of this Book, Google earth is now animating all water Imagery like a Disney Cartoon! They are also Clouding areas, even if you specify no weather. There are indications, since I started this Research in July, 2015, that Earth tones are being altered, and hazed-over.

I can only think of ONE reason for these combined actions: to Hide and Ob-fuscate. Judging by past history, this never works, it only makes people less-likely to believe in your Product, and, therefore less-likely to use it!

In--the-End, TRUTH always prevails, and if I can RAISE another fellow Ant, to the correct Altitude, then, perhaps, they, too...can SEE Mona Lisa's Enigmatic SMILE!

Fig 104

THE EARTH AS CREATED BY ANUNNAKI

19. TAOS ESSENCE

EVEN THE TAO, MUST HAVE A SPIRIT...

ESSENCE

A BRIGHT-CLEAR LIGHT

A PERFECT TRUTH

A PROFOUND JOY...

...NEVERENDING

...LOVE...

A TRUE POWER-

 THE ADDITION OF WHICH

MAKES US -MORE-

THAN THE SUM

OF OUR PARTS

NO GODS

NO DEMONS

NO JUSTICE...

JUST US

AND THE PATH,

A LIGHTED SPIRAL-STAIRCASE,

EACH TRAVELS

ALONE

WITH THE ULTIMATE

EVENTUAL GOAL

OF UNITY

Until The Promised Return,

 I Remain a Son Of The Sun

-Namaste

Dan Hoquist, although unable to attend College, due to financial reasons, had won a California State Scholarship after scoring in the top 3% on the SAT Test. Became a Seeker of Truths, through the 'school' of Experience, and the Intimacy of Personal Human Interaction, across the American Cultural Spectrum. Is a Journeyman Sheet Metal Worker, and former Carpenter, last Worked as a Mail Carrier (both Rural, and Clerk for the United States Postal Service). Disabled in 2005, was forced into Medical retirement, but found that 'Mind', still works Full-Time. This Book illustrates that work, and, even though Physically unable to do much 'real' Research, seeks now, to share the knowledge gained, in the hope that Publicity will help Protect the Archaeological Area that he discovered in Southwest Texas, also known as the Marfa Plateau, and to stir an increasingly 'Microscopic' Academia, to see the larger picture, and thus to 'Transparency', giving BACK the knowledge, for Humanity's future. Knowledge NOT shared, might as well be Knowledge Lost. And will Benefit none.

www.ingramcontent.com/pod-product-compliance
Lightning Source LLC
Chambersburg PA
CBHW040806200526
45159CB00022B/24